VIRUSES AS COMPLEX

ADAPTIVE SYSTEMS

PRIMERS IN COMPLEX SYSTEMS

VOLUMES PUBLISHED IN THE SERIES

Viruses as Complex Adaptive Systems,
by Ricard Solé and Santiago F. Elena (2019)

Natural Complexity: A Modeling Handbook,
by Paul Charbonneau (2017)

Spin Glasses and Complexity,
by Daniel L. Stein and Charles M. Newman (2013)

Diversity and Complexity,
by Scott E. Page (2011)

Phase Transitions,
by Ricard V. Solé (2011)

Ant Encounters: Interaction Networks and Colony Behavior,
by Deborah M. Gordon (2010)

VIRUSES AS COMPLEX ADAPTIVE SYSTEMS

Ricard Solé and Santiago F. Elena

PRINCETON UNIVERSITY PRESS
Princeton & Oxford

Published by Princeton University Press,
41 William Street, Princeton, New Jersey 08540
6 Oxford Street, Woodstock, Oxfordshire OX20 1TR

press.princeton.edu

ISBN 978-0-691-15884-6

LCCN 2018954901

British Library Cataloging-in-Publication Data is available

Editorial: Alison Kalett and Lauren Bucca
Production Editorial: Brigitte Pelner
Jacket/Cover Credit: Cover art courtesy of Shutterstock
Production: Jacqueline Poirier
Publicity: Alyssa Sanford

This book has been composed in Adobe Garamond and
Helvetica Neue
Printed on acid-free paper ∞

Typeset by Nova Techset Pvt Ltd, Bangalore, India
Printed in the United States of America

1 3 5 7 9 10 8 6 4 2

CONTENTS

PREFACE

Few examples of the way complexity unfolds in nature (and beyond it) are as fascinating as viruses. Viruses are hypothesized by some to predate the origins of life and its micro- and macroevolutionary play. They strongly influence energy flows in complex ecosystems. They maintain all kinds of relationships with their hosts, from mutualism to pure parasitism. They are responsible for some of the most deadly pandemics and yet have been coevolving with us throughout all our common history. They have played a major role in our understanding and manipulation of life and have also attracted the interest of biologists, physicists, and computer scientists alike.

Many key questions emerge when thinking of their nature and relevance: What exactly are they? Are they living entities? How did they originate? How similar are computer viruses to their living counterparts? How complex can they become? What is their role in shaping the evolution of complex organisms? What role have they played in the development of society? Can they be compared with software programs, to be run inside their hosts, who provide the appropriate hardware? Why are there so many new emergent viruses and how do they emerge? These questions will be addressed in this book.

Viruses are complex systems, spanning orders of magnitude in size, and an enormous variety of life cycles and habitats. But the study of their behavior and structure, particularly from interdisciplinary frameworks, has also revealed a number of universal patterns of organization. RNA-based viruses display high mutation rates, as predicted by theoretical developments from the 1970s. They live at the edge of disorder, where high instability but also adaptability occur. This edge is related to a phase transition phenomenon, and its presence is tied to the genetic nature of populations, often described as quasispecies. Other viruses, such as the mimiviruses, are so big that they are actually larger than some of the smallest cells we know. Their life cycle reveals surprising properties, placing them in a new position as a parasitic life form. We will discuss the potential boundaries of the viral morphospace and their importance, as well as theoretical models of viral population dynamics and self-assembly. Models and theoretical approximations have played a key role in this area, in particular the concept of fitness landscapes given their importance in defining the dynamics of their genetically complex populations.

Viruses have shaped the evolution of cells, organisms, ecosystems and even the biosphere. Such influence spans all scales of biological organization, from genomes to the planet (figure 1). Their dynamics involve nonlinear phenomena, tipping points and self-organization processes that have many commonalities with other biological and nonbiological systems. Such similarities might hide universal properties that pervade complexity and in this respect viruses offer a unique window into the origins of complex systems. In this context, we will also pay attention to other systems, such as computer viruses, that share unexpected similarities.

Although many excellent books exist concerning the population biology and ecology of viruses or the role they have played within the context of epidemics, there is no major contribution in

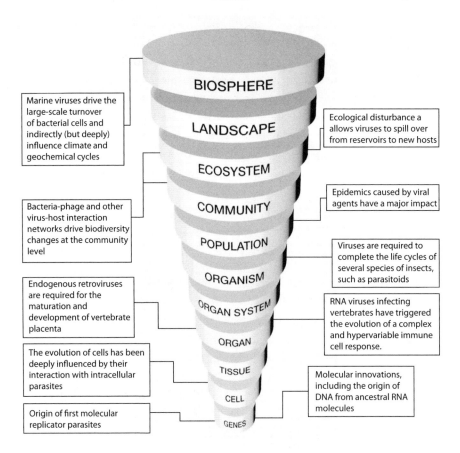

Figure 1. All scales of life are affected by and affect the diversity and evolution of viruses in the biosphere. In this diagram, some key processes and phenomena influenced or caused by viruses are indicated (central picture after Odum and Barrett (2005)).

the existing literature that provides the unifying picture presented here. The book has been written to be of interest to researchers and graduate students in virology, evolution, mathematical biology, and physics of complex systems. But we also hope it will be appealing to advanced undergraduate students interested in broad questions about viruses: where they come from, where they may go, and whether they are alive or not. Because of its multidisciplinary nature, some parts of the book require familiarity with intermediate-level algebra and calculus. Similarly, the book requires a basic understanding of molecular biology and the processes of information flow within the cells from DNA to proteins. We have made an effort to write every chapter in the most intelligible way, and more biology-oriented readers may skip over the mathematical details but still catch a good sense of what models represent and prove.

We would like to thank our many colleagues within virology and complex systems who have been helpful in shaping the ideas presented here, from both the experimental and theoretical sides, including some researchers, visitors, and good friends from the Santa Fe Institute. In particular, we would like to thank Lin Chao, Paqui de la Iglesia, Esteban Domingo, Stephanie Forrest, Fernando García-Arenal, Murray Gell-Mann, John J. Holland, Stuart Kauffman, Eugene Koonin, Chris Langton, Susanna Manrubia, Melanie Moses, Andrés Moya, Tom Ray, Rafael Sanjuán, Josep Sardanyés, Joan Saldanya, Peter Schuster, Paul Turner, Sergi Valverde, Eörs Szathmáry, Mark Zwart, plus a quite long list of PhD students, postdocs, technicians, and collaborators both in Barcelona and in València

The late physicist John Wheeler said about science: 'We live on an island surrounded by a sea of ignorance. As our island grows, so does the shore of our ignorance." This applies too to the particular island that defines today's understanding of viruses. Very often, new discoveries deeply change our perspective, expanding our understanding of the viral universe while the shoreline enlarges and new questions emerge. We hope this book will help the reader to walk safely across the tangled boundaries between the mainland and the uncharted waters.

VIRUSES AS COMPLEX

ADAPTIVE SYSTEMS

1

THE VIROSPHERE

1.1 Deep Microspace Field

The first image shown in this chapter (figure 1.1a), on the left side, is a deep field image from the far universe taken by the Hubble telescope. It was taken in 1995 and it covers an area of just about one 24-millionth of the whole sky, supposedly devoid of stars. And yet, the resulting picture, filled not with a few stars but many whole galaxies, takes your breath away. Almost all objects are galaxies that emerged in the early stages of the universe. The image next to it also recalls the deep universe. Is this picture an example of a cluster of stars or galaxies? Despite the similarities, the picture has been made using a special technique known as epifluorescence microphotography and deals with a vastly smaller scale: a small area of a drop of sea water. Even such a small amount of matter includes a huge number of microbial organisms. And here, too, scientists found much more than they would have ever suspected. The tiniest bright spots are viruses, followed in size by bacteria and archaea cells (medium-sized, around 0.5 μm in size) and also a few larger spots associated to protozoan organisms. These are no astronomic objects and yet an entire living universe inhabits the water drop.

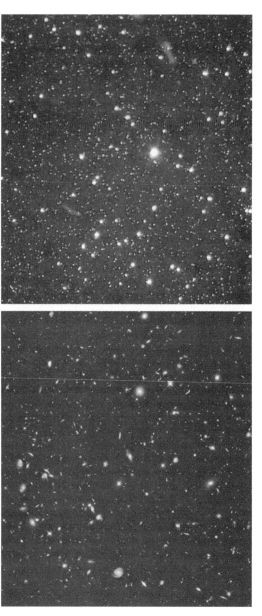

Figure 1.1. Large and small universes. The left picture is an image from the Hubble Deep Field Team and NASA. The right picture involves a very small, living system, namely the small-scale ecology contained in a sea drop of water, where marine viruses, bacteria, and protists appear as small, medium, and large bright spots, respectively. Adapted from Fuhrman (2009).

As it occurred with the discovery of the real dimensions of the universe, which was limited to our Milky Way until the use of powerful telescopes in the 1920s (allowing us to see that ours was just one galaxy among many), novel sequencing techniques uncovered a vast hidden microbial diversity in the oceans, from insects and other invertebrates and from non-cultivated plants, all of which were poorly sampled before. As more and more microbial diversity was uncovered, a feeling built in the community that it may be just the tip of the iceberg (DeLong 1997). The rise of metagenomics[1] revealed a completely unexpected, almost astronomic diversity of marine viruses. These viruses were virtually out of the ecological description of ocean life until then, but we know now that—like dark matter in our universe—this hidden part of the living biosphere turned out to be essential to actually understanding how the biosphere works. The presence of viral diversity and its importance is highlighted by the observation that every new sequenced virome includes new sequences (Angly et al. 2006; Kristensen et al. 2010; Simmons 2015; Roossink et al. 2015; Brum et al. 2015). It can be said that the perceived relevance of marine viruses (and the biosphere in general) has moved from anecdotal evidence to the fact that they are the most abundant biological entities, amounting to 10 to 10^2 viruses over cells.

These numbers are indeed impressive and since long before the discovery of marine viruses we have been dealing with these entities through our evolutionary history. As we will discuss in the following chapters, viruses have played a major role not only in our evolutionary but also in our recent history. They have been shaping our genomes and physiological features in surprising ways. They can be deadly or good. They can change

---

[1]Metagenomics consists of the characterization of genomic materials from environmental samples. By using advanced nucleid acid-sequencing techniques it is now possible to characterize microorganisms present in the samples without the need of isolating and cultivating them.

so fast and affect so many cellular pathways that mounting evidence supports a picture of evolution as largely dependent on the driving force provided by these entities. Their importance if so great that a "virocentric" perspective of evolution cannot be avoided (Koonin and Dolja 2013). Viruses as apparently innocent as those causing flu have killed millions of humans (and many other species), becoming a threat to our survival. But without them many major evolutionary events would have never happened.

Some viruses' names have become popular in the media because of their terrifying impact through deadly pandemic events. Two of them in particular are on top of our list: the *Human immunodeficiency virus type 1* (HIV-1, figure 1.2) and the Ebola virus (EBOV, figure 1.3). They both illustrate the simplicity that can be achieved by viral agents, equipped in most cases by small genomes where a few genes coding for the essential components and copying machinery often overlap in order to maximize information compression. Both examples involve small structures, high mutation rates and a huge capacity to trigger fear. But they differ in many ways. The molecular logic of their replication, their repertoire of cellular targets to infect, their origins, and how they spread through human populations are different.

HIV-1 became known as a real threat once new cases of a previously unknown disease started to become common in the 1980s. No one would have suspected then that the new pathogen would spread around the entire planet and produce a great pandemic killing tens of millions of individuals. It became known at some point that the virus was hiding inside cells and that carriers were free of symptoms over a long time interval before a collapse of the immune system occurred with fatal consequences. During this silent period, infections would occur and the virus propagated exponentially. HIV-1 spread through both rich and poor countries. Until its biology was well understood and its

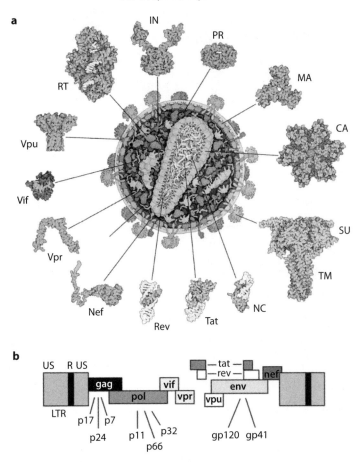

Figure 1.2. A well-known example of an RNA virus that has been responsible for one of the worst pandemics ever: HIV-1. In (a) we display the basic structure of the whole virus (central figure), and all the key molecular components are also indicated (image adapted from Goodsell (2012)). In (b) the HIV-1 small genome is schematically shown, involving just three structural genes (*gag*, *pol*, and *env*) along with regulatory elements.

Achilles' heel was discovered, the death toll grew over time. Only scientific and clinical research, including a great deal of modeling effort, eventually enabled us to properly fight back.

a

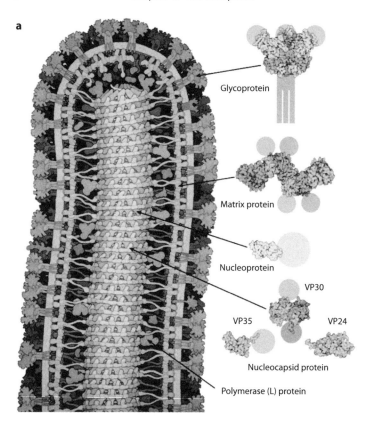

Glycoprotein

Matrix protein

Nucleoprotein

VP30

VP35   VP24

Nucleocapsid protein

Polymerase (L) protein

b

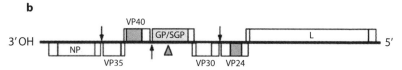

3′OH   5′

Figure 1.3. EBOV (a) an emergent virus that has likely jumped from bats
to great apes and humans. It belongs to the family of filoviruses and
has a characteristic filamentous shape. The key molecular components
are also displayed (image adapted from Goodsell (2012)). (b) Schematic
representation of the genome, encoding for seven structural and one
nonstructural protein.

EBOV, on the other hand, represents a good example of another emergent pathogen notable for the bloody and deadly way in which it interacts with the human host. But in this case, the rapid damage caused to the patients prevents the virus from spreading on a global scale. However, poverty, reduced investment in healthcare, and some cultural factors have to be blamed for most EBOV outbreaks.

1.2 The Expanding Viral Universe

The impact of this hidden *virosphere* on ecosystem functioning can be summarized by means of some basic numbers (Suttle 2005; Weitz 2017). The number of viruses that might be present in the entire marine biota is 10^{30} (a 1 followed by 30 zeros) and the number of infection events taking place every second would amount to no less than 10^{23}. As a consequence of infections, viruses kill around 20% of the total microbial biomass in a single day, thus forcing a constant and large-scale population turnover. Since the microbial component of the marine biota is responsible for a major fraction of energy flows, the obvious consequence is that large-scale ecological processes are strongly constrained by the viral component of the biosphere. Marine viruses illustrate one of the most obvious results of ecological research: the realization that our planet is dominated by microbes and, very especially, by viruses.

In figure 1.4 we summarize this dominance using two main quantitative measures: total biomass and population abundance in marine communities. The biomass is clearly dominated by bacteria, with prokaryotes and viruses following closely in small fractions. However, the total number of individuals clearly differs from what is represented in the biomass picture (figure 1.4 left). Here viruses greatly outnumber other taxa, consistently with our previous picture. As pointed out by Koonin and Dolja (2013), it

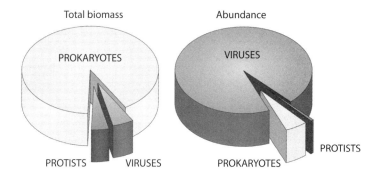

Figure 1.4. Viruses are by far the most abundant biological entities in the oceans, comprising approximately 94% of the nucleic acid-containing particles, but they only amount for 5% of the total biomass. By contrast, even though prokaryotes represent less than 10% of the nucleic acid-containing particles, they represent more than 90% of the biomass (diagram adapted from Suttle (2007)).

can be claimed that the water in the ocean is literally a virus soup with up to 10^9 viral particles per milliliter.

A very different but not less rich facet of the viral universe plays a crucial role in our own bodies. It is well known that a human needs to be seen not as an isolated entity carrying around 20,000 genes, but instead as a complex consortium of species. In particular, we are the carriers of a vast ecological web of interactions that take place among the many species of microorganisms that colonize our mouth, lungs, gut, or skin. This is known as the *microbiome*. The microbial part of ourselves carries around three million additional genes and has been coevolving with us for millions of years (Boulang and Nagler 2016; Wesemann and Nagler 2016; Taur and Pamer 2016). After the recognition of the major impact of the microbial part of our nature, the so-called *virome*, a no less interesting problem has to do with the inevitable role played by the microbe's parasites (Minot et al. 2016).

The example used above provides just a first glimpse of the enormous relevance of viruses. In this chapter, we will provide an overview of the complexity of the virosphere, addressing several key questions: What is this virosphere made of? How have viruses so successfully expanded over every single scale from bacteria to humans and even to other viruses? Is this virosphere very diverse? These questions will help to define the vast scenario that we plan to explore in this book. It is not only an extraordinary example of our biosphere's complexity; viruses themselves are a rich, and sometimes unexpected, instance of complex systems that perfectly illustrates the tempo and mode of complexity evolution and how it pervades nature.

1.3 Structural and Genetic Diversity

Viruses inhabit a domain of size ranges that spans the broad interval between molecular structures and cells. Some viruses are so small that it took a long time to detect them (figure 1.5). They were first reported in 1892 by Dmitri Ivanovsky, a Russian scientist who was studying the process of transmission of a tobacco disease. Some unknown pathogen was damaging the plant tissues and in an experiment he filtered a suspension of infected tissues through a ceramic filter, which was known to retain bacterial cells. Once filtered, the suspension was free of bacteria and yet capable of infecting, indicating that a smaller class of biological agent was responsible for the disease. In this way, the first virus was discovered: *Tobacco mosaic virus* (TMV). Other scientists independently confirmed the existence of this new class of entities, and the invention of the electron microscope was, along with the development of molecular genetics, an especially important breakthrough, since those invisible agents became visible and their internal structures and genetic components became available to study.

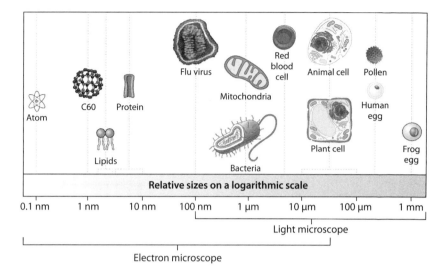

Figure 1.5. Viruses occupy an intermediate position between the macro-molecules and living cells and organelles. Most of them are 10^2 times smaller than the smallest cells, though the discovery of many giant viruses along the last decade has changed this view.

Because of their simplicity, viruses cannot replicate outside of the cellular context. They need the cell machinery to make copies of themselves (see chapter 2), and that of course makes a big difference. What is perhaps most impressive of viruses is that they are paramount examples of diversity in all kinds of contexts. They embody a vast range of replication strategies and structural forms of organization, spanning orders of magnitude in genome size and complexity. At the lowest extreme of the complexity continuum are the viroids, which are small folded RNA chains no more than a couple of hundred nucleotides long that do not encode any protein (Flores et al. 2014). At the other end of the complexity are viruses so large that they were initially mistaken for bacteria. This group includes mimiviruses, iridoviruses, pithoviruses, pandoraviruses, and other members of the brotherhood of giant viruses. In figure 1.6 we show a few

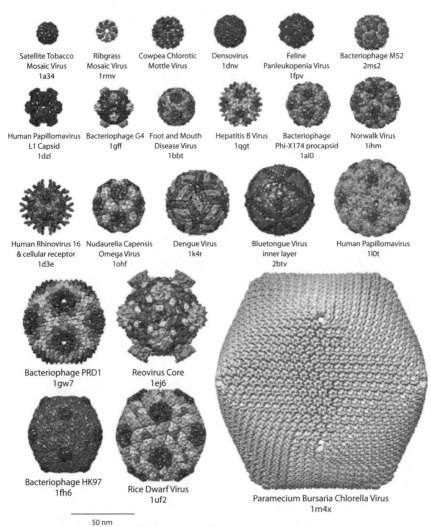

Satellite Tobacco Mosaic Virus 1a34

Ribgrass Mosaic Virus 1rmv

Cowpea Chlorotic Mottle Virus

Densovirus 1dnv

Feline Panleukopenia Virus 1fpv

Bacteriophage M52 2ms2

Human Papillomavirus L1 Capsid 1dzl

Bacteriophage G4 1gff

Foot and Mouth Disease Virus 1bbt

Hepatitis B Virus 1qgt

Bacteriophage Phi-X174 procapsid 1al0

Norwalk Virus 1ihm

Human Rhinovirus 16 & cellular receptor 1d3e

Nudaurelia Capensis Omega Virus 1ohf

Dengue Virus 1k4r

Bluetongue Virus inner layer 2btv

Human Papillomavirus 1l0t

Bacteriophage PRD1 1gw7

Reovirus Core 1ej6

Bacteriophage HK97 1fh6

Rice Dwarf Virus 1uf2

Paramecium Bursaria Chlorella Virus 1m4x

50 nm

Figure 1.6. Some examples of regular structures found in viruses, from very small, such as $\phi X174$, whose genome was the first to be sequenced, to the largest known mimiviruses, which involve hundreds of genes and have a size even larger than that of the smallest bacteria.

examples of some well-known viruses so we can appreciate the broad range of sizes. The smaller viruses include the famous $\phi X174$, the first entity whose genome (a circular, single-stranded DNA molecule) was fully sequenced (Sanger et al. 1977). The genome of this bacteriophage involves just 5,386 nucleotides, required to encode 11 proteins. But we can also find smaller viruses with some interesting traits besides their tiny size. This is the case of the satellite RNA of TMV, with a 1,063 single-stranded RNA genome which codes just for the capsid and one other protein. This satellite infects tobacco plants already infected with TMV, worsening their symptoms. In this case the satellite virus (that is why this name) needs the cell machinery both of the plant *and* the one from its host virus, TMV.

At the other extreme of the size spectrum, we have a member of the *Mimiviridae* family, including the largest known viruses. The first microscope observations (in 1992) found them infecting amoebas, and given their large size and staining properties they were assumed to be gram-positive bacteria. A correct identification of these microorganisms as true viruses took place eleven years later (La Scola et al. 2003). Since then, many other types have been found (Abergel et al. 2015). Their genome size is comparable with that of cellular genomes, and can be longer than one million base pairs. The finding of this group (as will be discussed in chapter 7) created novel views of the boundaries between living and nonliving entities.

An especially remarkable feature of viruses is their enormous genetic diversity. This diversity is not just a matter of size and composition: it is about the logic of the replication and its evolutionary consequences for the rest of life on earth. There is a striking contrast between the homogeneous nature of information processing that takes place in the nonviral world and what occurs in the virosphere. Cellular genomes replicate thanks to a highly complex molecular machinery based on the transcription of a double-stranded DNA molecule into an RNA chain that is single-stranded, which itself is then translated by another equally giant molecular complex (the ribosomes) into the proteins necessary to build the whole replication complex (Crick 1970). All cellular organisms respond to this pattern, with very rare deviations. In the virosphere, by contrast, *all* kinds of RNA and DNA combinations and interconversions among them are observable. Such a broad spectrum of genetic strategies allows for a potential evolution that makes viruses a true "genomic laboratory" (Koonin and Dolja 2013). Indeed, one of the first attempts to classify viruses into groups with similar properties was by David Baltimore (1971), based on the type of genetic material (either DNA or RNA, single- or double-stranded) and replication strategy. According to Baltimore's scheme seven groups of viruses can be defined (figure 1.7):

1. Group I is formed by those viruses having a double-stranded (ds) DNA genome. They usually replicate in the nuclei of infected cells and use cellular proteins for their replication. Examples are the herpes viruses and the smallpox virus.
2. Group II includes all viruses having a single-stranded (ss) DNA genome. They also use the cellular machinery for their replication. Examples are the Canine parvovirus and the plant geminiviruses.
3. Group III have dsRNA genomes and replicate in the cytoplasm of the infected cells. They encode for their own replication enzymes. Examples are some fungal viruses.
4. Groups IV and V are the most abundant classes and have genomes of ssRNA of either positive sense (group IV) or negative sense (group V). Positive sense means that the molecule encapsidated can directly be translated by the cellular translation machinery, whereas negative sense means that the molecule

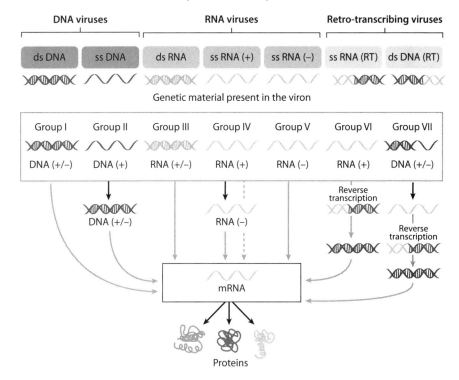

Figure 1.7. A simplified schematic representation of Baltimore's classi-fication of viruses according to the nature of their genomes and their replication intermediates. Adapted from Flint et al. (2015).

encapsidated has to be first transcribed into its complement and then can be translated into proteins by the cellular ribosomes. Most known viruses belong to one of these two families: TMV, *Hepatitis C virus* (HCV), *Foot-and-mouth disease virus*, EBOV, *Yellow fever virus*, and the several influenza viruses.

5. Group VI corresponds to viruses having a positive sense ssRNA genome that is replicated via an intermediate DNA molecule. This group corresponds to the well-known retroviruses whose most characteristic

representative is the HIV-1. All retroviruses encode for an enzyme, the reserve transcriptase, that synthetizes DNA using RNA as template.

6. Group VII corresponds to dsDNA viruses that replicate through an ssRNA intermediate. This small group of viruses, whose representative is *Hepatitis B virus* (HBV), also encodes for a retrotranscriptase (Baltimore 1971; Flint et al. 2015).

No less important here is the fact that viruses have coevolved with their hosts since life began on our planet. This is particularly obvious from the study and sequencing of genomes of plants and vertebrates, which display large amounts of virus-related sequences (Aiewsakun and Katzourakis 2015; Ryan 2016; Mushegian and Elena 2015).

1.4 Viral Planet

Since their discovery, the importance of viruses and the understanding of their ecological and evolutionary impact have only been growing over the last century. It was revealed early that they are responsible for many human diseases, and their genetic plasticity and easy manipulation were crucial in the early days of molecular biology, when bacteriophages (viruses infecting bacteria) were used to test many fundamental ideas concerning the nature of heritable information and the genetic code (Morange 2000; Creager 2002; Cairns et al. 2007). As mentioned above, it was later found that they have a great impact in the ecology of marine ecosystems, with major consequences not only for populations, but for the planet as a whole (see below).

Our biosphere is a complex adaptive system (Levin 1998) where multiple scales of organization are shaped by a number of physical, developmental, ecological, and historical factors. In all these scales viruses play a relevant role. Matter and energy flows take place through tangled networks of interacting species. A vast range of biomasses are involved, from the largest animals to the smallest cells. But every single organism has at least one, if not many, virus associated to it. Both unicellular and multicellular life forms are rich niches where a virus can find opportunities to evolve. The interactions among hosts and viruses are not always parasitic (see chapter 4) and often lead to disease outbreaks that can have great consequences (discussed in chapter 5). Getting back to the oceans, since viruses have a great impact in the population dynamics of their hosts, and indirect impact on other organisms that prey on their hosts, they also affect deeply the large-scale dynamics of nutrient cycles (Weitz 2016).

Figure 1.8 summarizes the magnitude of the impact of viruses on carbon cycling in the oceans. For comparison, figure 1.8a displays the effects of anthropogenic activities associated to the intensive use of fossil fuels along with the role played by land forests. The basic network of carbon flows[2] is described in figure 1.8b. Viruses kill plankton cells at a rapid pace, leading to both particulate organic carbon (POC) and dissolved organic carbon (DOC). Instead of sinking to the depths, where huge amounts of carbon are being stored, both become suspended or dissolved and thus do not sink. Without the presence of viruses, a large fraction of planktonic carbon is retained in surface waters where it can respire and be photo-oxidized and in chemical equilibrium with the atmosphere. The lytic infection triggered by viruses results in further viral particles and in a complex mixture of molecular pieces forming the cellular debris. This includes small molecules (both monomers and polymers), colloids, and cellular fragments. Most of these components will be incorporated into bacteria and other organisms living in the upper ocean layer (Fuhrman 1999).

---

[2]Carbon flows are given in Gigatonnes (Gt) per year. A gigatonne indicates one billion tonnes or 10^{15} grams.

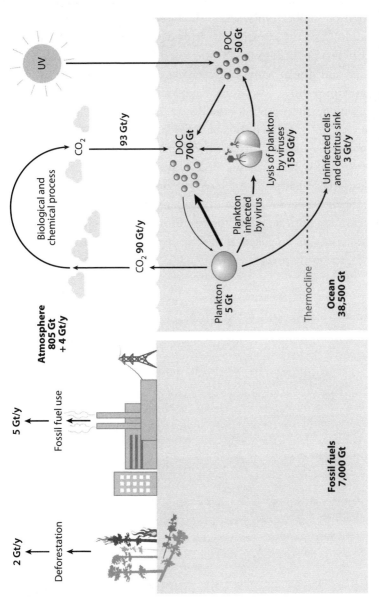

Figure 1.8. Viruses have a deep impact on the carbon cycling at a planetary scale. Because of their effects on the death of plankton, the resulting particulate organic carbon (POC) from cell lysis remains close to the water surface (instead of just sinking into the deep ocean). This strongly affects he balances of CO_2 in the atmosphere.

Viruses define in some ways the coastline of life; little can be understood about evolution of the biosphere without taking viruses into account as major (perhaps dominant) players. It is often said that life is almost everywhere in the planet except inside volcanoes and similar hyper-extreme environments. We can also say that viruses thrive everywhere where life has flourished. This might well be the case for *any* life we find in other worlds, with molecular parasites inevitably associated to self-replicating autonomous entities. In this book we aim to explore the origins of their special status, their universal features and origins from a complex systems perspective. Moreover, viruses are not confined to life. Their properties and propagation dynamics have been an inspiration for understanding the rise of some key evolutionary novelties, such as language and other aspects of cultural and technological evolution.

2

ALIVE OR DEAD?

2.1 Computation and Life

Molecular biology and information technology (IT) emerged almost simultaneously around the middle of the twentieth century and have evolved in parallel since then. Despite the great differences existing between living structures and computer hardware and software, a continuous exchange of ideas and terms took place in the early development of both disciplines (Maynard Smith 2001). An interesting convergence also took place. Engineers building the new technological engines capable of manipulating information used previous theoretical models of computation (as defined by Alan Turing), but they also actively contributed to another important (and much older) domain: the coding and decoding of messages.

Around the 1950s, coding and decoding secret information became a major target of the Cold War efforts. Computer designers and programmers had also to find ways of performing computations at the lowest cost. The early machines were still expensive and had a limited power, and everything needed to be properly designed under strong constraints. That meant writing short, optimized programs, using appropriate coding schemes, and compressing information. A struggle that was being performed by small living beings already for billions of years of evolution, although such a race was unnoticed to any scientist by the time the IT revolution started. The reason we mention these similarities between the two fields is that they require the presence of machines (*in silico* or *in vivo*) capable of executing programs. These two components of computation are usually known as hardware and software, and for understanding the nature of viruses we need to approach the question of how close they are to containing these two components.

The classical view of viruses, seen as intracellular parasites requiring the available molecular machinery to replicate themselves, suggests that we can consider them as some sort of program. In this context, viruses would mainly be considered as encapsulated pieces of software, to be executed by the cellular host hardware. In all these cases, the function that is being executed is a *computation*, and using this concept will be extremely useful in our exploration of viruses as computational objects. The idea of a general machine capable of performing computations was formalized in theoretical terms by the British mathematician Alan Turing.[1] In figure 2.1 we display a mechanical implementation of Turing's view of a machine performing computations (Hopcroft 1984; Bennett and Landauer 1985). Here a computer is represented by a printing head moving along a long tape where symbols are written and can be read and typed by the machine. Despite the simplicity of this *Turing machine*, it can be shown that (given enough time) it can perform any computation executable by any digital computer.

An interesting and often unnoticed implication of Turing's result has to do with the molecular counterparts of the Turing

---

[1] Turing's goal was not to define computing machines but instead to address a major problem posed by David Hilbert concerning a way to define a systematic method to establish the truth or falsity of any mathematical statement. The turing machine's formulation of the method is a proof that no such method exists.

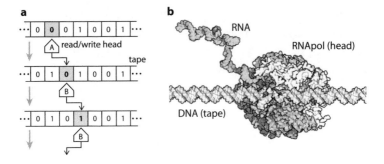

Figure 2.1. The Turing machine (a) is the abstract representation of a computational device that can be built in such a way it can perform any operation done by a computer. It is defined in terms of an infinite tape where 0's and 1's are written and that can be read and written on by a "head." This head moves along the tape, changing its internal state in a deterministic way. Several molecular processes within cells (b) remind us of this scheme. Here the RNA polymerase is shown "reading" the DNA chain and "writing" an RNA molecule as a result (adapted from Goodsell (2012)).

machine. Usually, the Turing approach to computation is only connected with computer viruses (see chapter 8), with no special attention to real viruses. When Turing formulated his theory, molecular biology did not exist and no one knew that molecular information was in fact stored in long (sometimes very long) polymer chains. Moreover, the process of transcription and translation strongly resembles the Turing metaphor (figure 2.1b). Such a match between a purely mathematical approach (Turing's machine does not pretend to entail a real computer) with the molecular logic of information processing and molecular replication is remarkable. It points toward a universal form of performing computations that might be inevitable when physical polymers are at work.

If cells include (among other things) these molecular machines capable of identifying and reading biological polymers, it seems clear that viral genomes can be seen as pieces of tape with start and stop signals and that they will be read and interpreted by the cell polymerases and ribosomes, which share a common alphabet. Using this framework, we can separate cellular computations from viral information by assuming that cells include both heads and tapes acting on a very large set of messages, whereas viruses deal only with tapes, lacking—as machines parasitizing machines—specific machinery for dealing with the read-write process. Is this picture correct? Digging for an answer requires looking at viruses from different perspectives.

2.2 Viruses as Replicating Machines

Molecular biology enriched its emerging lexicon from IT technology. Terms such as coding and decoding, transcription and translation, proofreading, mismatch, redundancy, synonymous, messenger, or even library, are fully incorporated within the current jargon of modern biology. Soon, it also became clear that some major differences were at work. A strong convergence between computational and biological systems was foreshadowed by the work done by the great mathematician John von Neumann, a major player in shaping the computer revolution. Among other fundamental problems linking biological and technological systems, von Neumann was interested in the possibility that a machine could self-replicate in some autonomous way (von Neumann 1966). What would such a machine look like? What would be the minimal logic requirements for self-reproduction? This is a relevant question in our context. Self-replication is a key feature of living beings and is of obvious relevance for our treatment of viruses.

The theoretical basis of self-replication[2] was approached by von Neumann using a very abstract (but also general) view,

---

[2]It is worth mentioning that von Neumann was a mathematician, and his work, completed around 1956, was years ahead of the discovery made by James Watson, Francis Crick, and Rosalind Franklin of DNA structure and the implicit molecular logic of genetic inheritance.

ignoring the exact nature of the components and the specific functions executed by the machine. One of the great advantages of von Neumann was his familiarity with hardware (due to his involvement in the design of ENIAC),[3] especially because his previous work on a rather novel—and revolutionary—aspect of information technology: software (Dyson 2012). Von Neumann understood the importance of information and its relevance in providing a set of instructions to replicate the whole machine while replicating the instructions themselves, a set of requirements that defined the minimal requirements, as summarized in figure 2.2. Here we can see that the system is composed by three parts labelled A, B, and C, corresponding to the constructor, the duplicator, and the controller, respectively. Crucially, a blueprint of instructions $\phi(A, B, C)$ is required to provide the set of instructions needed to make copies of the machine.

Today we know very well that this is also the logic of living cells: DNA operates as the basic set of instructions $\phi(A + B + C)$ defining the hardware, while the machinery, A, required to replicate the instructions in an accurate manner (defined by a highly complex network of proteins) is provided by a set of molecular machines, such as polymerases, which interact with DNA and replicating the instructions when the cell divides. Ribosomes that interpret the information transcribed from DNA into RNA correspond to the duplicator B. As indicated in the scheme of figure 2.2 the cell is defined by two parts that interact. What would be the diagram associated to viral replication? This scheme can be easily defined by a piece of software ϕ_v that requires the cell machinery to make a replica of itself (figure 2.2, left). In this scheme, the virus blueprint instruction ϕ_v uses the cell machinery (A and B) to generate copies of itself, which are then ready for a new round. In this minimal setting (where the virus is considered just a naked replicator) the only operation

---

[3] Electronic Numerical Integrator and Computer.

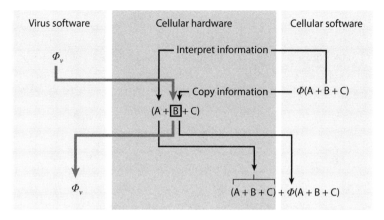

Figure 2.2. John von Neumann found the general conditions for a self-reproducing system. In 1956 he found that this abstract problem had a general solution, schematically described in the diagram. A given machine with the potential for replicating itself should include three basic elements, namely: a *constructor A*, a *duplicator B*, and a *controller C* and a set of instructions $\phi(A + B + C)$ that specify how to construct the machine parts A, B, and C. In this sense, A, B, and C are the hardware (central area), while $\phi(A, B, C)$ corresponds to the software. A virus, in its simplest form, can be understood as a piece of software that uses the cell replication machinery to produce a new copy of viral information (thin gray lines). Adapted from Shirt-Eddiss (2016).

needed uses the molecular "head" capable of reading the viral chain and creates additional copies of it. In contrast with the full scheme associated to cell replication, just a molecular parasite needs to be copied.

This is of course an oversimplified picture. Even the smallest viruses are typically packed and need to infect their hosts, which in some way or another implies an evolved recognition process. The molecular parasite must be capable of detecting and attaching to its host, and the subsequent chain of events dragging the virus genome inside the cell are describable as some kind of information processing (thus representable in terms of a computation). Leaving aside these features, the previous scheme suggests that, if applicable, viruses are less than self-replicating machines, since they are actually dependent upon the operations executed by the machines they infect. Is this a formal argument against viruses as living systems?

2.3 Viruses as Phases of Matter

In chapter 1 we presented an overview of the diversity of known viruses. We paid attention to the differences in organization, size, ecological adaptation, and lifestyle. Moreover, in the previous sections we considered the nature of viral replication and virus life cycles from a complex systems, logical perspective. As it happens with their life forms, target hosts, and strategies of infection, the ways viruses reproduce span a wide array of possibilities. At the level of both structure and function, molecular parasites seem to display a great variety of designs. Indeed, the viral universe is vastly diverse and changing, but there are also some key, generic mechanisms associated to virus development that reveal deep and common physical processes pervading viral complexity. Similarly, there are strong constraints to structural organization in a large class of spherical viruses that remind us of the fundamental role played by physics. All these features are connected with the hardware of the system, and by inspecting its nature we can also better locate viruses within the abstract computational schemes outlined above.

Typically, the size of a given viral particle is proportional to the size of its genome. But it was early well known that the genome was much smaller, in terms of mass, than the total mass of the whole viral particle. This observation prompted Watson and Crick in particular to suggest that the viral capsid should be composed by the association of multiple copies of basic capsid protein(s). As noted by these authors (Crick and Watson 1956, 1957), the capsids of many viruses are formed from a minimum number of gene products, given the small size of viral genomes. As a consequence, spherical viruses should

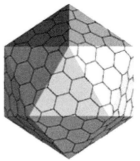

Figure 2.3. Packing a closed volume by using simple geometrical motifs. The left image shows five so-called *Platonic* shapes, obtained by using an identical motif (triangles, squares, pentagons, etc.) and packing them so that they minimize the resulting surface. Spherical viruses almost universally pack their genetic material by using icosaedra (right) through a process of self-assembly of multiple capsomers.

have the symmetry of regular polyhedra, also known as *Platonic solids* (figure 2.3). In these objects, all faces are identical perfect polygons that correspond to protein units. An icosahedron would be formed by 60 of such units, in which all protein units sit in identical environments; the largest shell of this kind is an icosahedron consisting of 60 equivalent subunits. Subsequent capsid structure determinations confirmed the special role of icosahedral symmetry, but also indicated that larger numbers of protein subunits were involved.

Spherical viruses and rod-like structures are two major classes of generic forms associated to mature virions infecting all types of hosts, from bacteria to higher plants and mammals. All these viruses are made by a tightly packed, regular distribution of a limited number of subunits forming capsids. The process of capsid assembly involves geometry, since regular packing needs some geometric constraints at the level of units. The fact that no virus carries out metabolic activity by itself indicated that, unlike cells, their assembly could be understood in terms of standard equilibrium thermodynamics. An elegant confirmation of this idea was the discovery that under *in vitro* conditions the rod-like TMV self-assembles spontaneously and unassisted into fully infectious viral particles (Frankel 1955). The basic sequence is summarized in figure 2.3a, where the self-assembly process is displayed. One single monomer unit gets polymerized into a helical structure, and this is a thermodynamically favorable process. Here the helical structure is an intrinsic property of the protein, which gets organized into disks (Klug 1999). The RNA gets attached to the growing rod as time proceeds. Regularities in the pattern of TMV genomic RNA folding into secondary structures serve as packaging signals, repeated according to capsid symmetry, aid formation of the required capsid protein conformers at defined positions, resulting in significantly enhanced assembly efficiency. The precise mechanistic roles of packaging signal interactions may vary between viruses (Stockley et al. 2013).

The widespread relevance of self-assembly mechanisms in orchestrating the causal process behind virus morphogenesis has an interesting consequence. It depends on energy-driven, physical forces, which have no direct connection with genetics or biology. They can be directly associated to physical laws, and thus we can treat viral assembly as a physics problem that might be more complex than a simple physics problem involving point particles and orbits, but it is in the end no different from the point of view of molecular forces and energy minimization processes (Rossman and Rao 2012).

How can such model be formulated and what are its consequences? Several studies have used well-defined physical models of capsomer structure and capsomer interactions, usually based on different extensions of well-known reaction kinetic models including self-assembly and polymerization (Kushner 1969), and in particular physical models of viral assembly based on the minimization of energy functions (Bruinsma et al. 2003; Zandi et al. 2004; Hagan 2014). Among the most interesting results

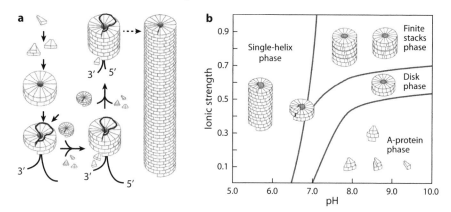

Figure 2.4. Self-assembly and growth of the TMV particle. The basic monomers required to build the entire virus particle self-assemble, forming a characteristic ordered, helical structure. The RNA chain attaches to this growing structure. The picture displays a phase diagram showing the ranges over which particular structural patterns of protein assemblies are stable (redrawn from Klug (1999)). Here two parameters are used, namely the pH and the ionic strength.

offered by these theories when applied to spherical viruses is the finding that the limited repertoire of possible icosahedral "solutions" corresponds to the minima of an energy landscape (Bruinsma et al. 2003), showing that physics pervades the constraints associated to the universe of viral forms. One important side effect of this result is the explanation for the discrete nature of possible icosahedral viruses and their "mathematical" nature (Stewart 1999).

The simplest picture of self-assembly processes is provided by a kinetic model leading to a dynamical pattern of aggregation characterized by the presence of a phase transition behavior (Dill and Bromberg 2011). This can be illustrated by a minimal model where a set of n monomers[4] aggregates into an aggregate. If we

---

[4]We use this model to discuss the self-assembly of viruses, but it has been used in many other contexts, particularly in the formation of micelles (Mouritsen 2005).

indicate by A_1 a single building block and by A_n the assembled system, the reaction for the whole process would be described by

$$nA_1 \xrightarrow{K} A_n, \qquad (2.1)$$

where $K = [A_n]/[A_1]^n$ is the associated equilibrium constant. If we indicate by C_0 the initial concentration of A_1, the reaction kinetics for A_1 would be described by a nonlinear equation $dA_1/dt = -KA_1^n$ whose solution is given by

$$A_1(t) = \left[\frac{C_0}{1 + KC_0(n-1)t} \right]^{\frac{1}{n-1}}, \qquad (2.2)$$

which can be shown to display two markedly different behaviors as a function of the equilibrium parameter. This can be done by looking at the fraction $v(x)$ of components associated to assembled aggregates as a function of K, where $x = [A_1]$. Since $[A_1] + n[A_n]$ is the total number of monomers and $[A_1] + [A_n]$ is the total number of "objects" (either aggregates or single monomers), it can be shown that

$$v(x) = \frac{1 + nKx^{n-1}}{1 + Kx^{n-1}}, \qquad (2.3)$$

which exhibits a sharp transition close to a critical value

$$x_c = K^{-1/(n-1)} \qquad (2.4)$$

jumping from 1 to n (i.e., from all A_1s single to all in aggregate).

This cooperative behavior indicates that, once a given concentration of monomers is reached, the system experiences a rapid (and irreversible) transition into large structures with a characteristic size. This occurs in a thermodynamically favored direction, and thus the resulting assemblies are highly stable structures. In a chemical system formed by inert molecules, self-assembly takes place by an energy-minimization process (that is captured in the irreversible reaction described above), eventually ending up in stable assemblies. An infective virus particle will create the conditions for this transition as more and more copies accumulate until the right conditions for assembly are met. It can be said that the final part of the developmental path followed by a spherical virus life cycle requires a cooperative transition.

The specific application of this approach to the self-assembly of viral capsids can be used to obtain an expression for the fraction of capsids $f(\rho)$ present in a given system (such as the inner space of the cell) as a function of the monomer concentration (ρ), assuming that (as before) we neglect all molecular intermediates except free subunits (Hagan 2014). If we the total concentration of capsid units ρ_T is given by

$$\rho_T = \rho_1 + N\rho_N \tag{2.5}$$

where N is the number of capsomers in one capsid, and ρ_1 and ρ_N are the densities of single subunits and whole capsids, respectively, define the fraction of subunits in capsids as

$$f_c = \frac{N\rho_N}{\rho_T}; \tag{2.6}$$

a critical concentration ρ^* exists such that a phase transition occurs, separating a sub-critical phase with essentially no self-assembly of viral particles, i.e.,

$$f_c(\rho_T) \approx \left(\frac{\rho_T}{\rho^*}\right)^N \tag{2.7}$$

for $\rho_T \ll \rho^*$, and another phase where virus capsids form, and f reads

$$f_c(\rho_T) = 1 - \frac{\rho^*}{\rho_T} \tag{2.8}$$

when $\rho_T \gg \rho^*$. The phase transition curves predicted from the model are displayed in figure 2.5a, using three different values of N. An experimental test of this model is shown in figure 2.5b, where different concentrations of capsid subunits have been used under variable salt concentrations (which enhance the self-assembly process).

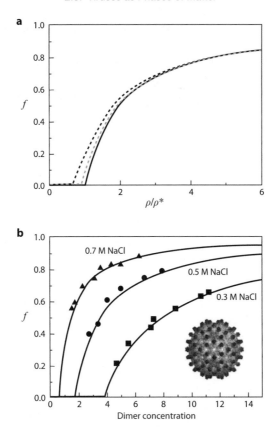

Figure 2.5. Phase transition in the capsid assembly process. In (a) we show the predicted theoretical result $f_c(\rho_T) = 1 - (\rho^*/\rho_T)$ for (from left to right) $N = 12, 60$, and $1,000$, respectively. $f(\rho)$ represents the fraction of capsids present in a given system as a function of the monomer concentration (ρ). An experimental test of this theoretical result is provided in figure (b) using empty capsids of HBV (inset) under different dimer subunit and salt concentrations. Adapted from Hagan (2014).

The need for self-assembly, in both cells and viruses, implies that part of von Neumann's scheme described above needs to be mapped into an embodied set of rules that cannot be separated from the self-organized nature of living matter. This is indicated

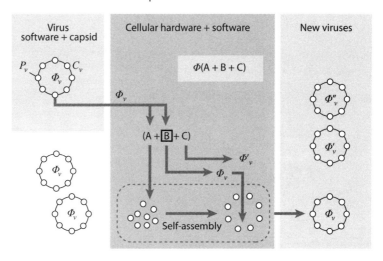

Figure 2.6. The extended von Neumann scheme for viral replication when both capsid and genome are considered. Here a virus is defined by a blueprint ϕ_v packed within a closed boundary C_v made of a given number of identical proteins P_v (left). Once it infects the cell, the viral information makes use of both A and B to build capsid proteins and new ϕ_v. Instead of a controlled process of development, viral packing involves self-assembly (dashed box) leading to a packed capsid containing a copy of the original genome, although mutated versions (ϕ', ϕ'') are also possible.

in the extended von Neumann diagram shown in figure 2.6. The two main differences (in relation to the scheme of figure 2.2) are the existence of a physical-driven mechanism of organization that is not encapsulated by the "program" given by ϕ and the fact that the outcome of the process can include a different blueprint as a result of mutations. What other factors can modify the initial scheme? Since viruses can be more complex and interact with the three components of the cellular machinery (A,B,C), and even themselves carry pieces of molecular machinery, the resulting interaction scheme can be far from a simple cell versus virus mixture.

These results provide useful insight into the ways some viruses exploit the self-organization properties of matter to create order out of chemical homogeneity.[5] These results suggest that we can, to some extent, look at viruses as molecular entities with some special capabilities associated to reproduction. But there is much more. In the next two sections we will consider first a classical experiment where such prevalence of chemical self-replication is the final outcome of artificial evolution. Secondly, we will look at the space of replicators to see if this simple image of viruses is correct.

2.4 Evolving Genome Reduction

An essential part of genome evolution has to do with the expansion and shrinking of genomes. In simple life forms, genome complexity correlates with genome length, since the information content stored in the genome needs to be made compact in order to compress all required instructions while replicating in an efficient way. For viruses, if more complex sets of decisions are required to infect their hosts, replicate, assemble, move out, or integrate, larger genomes are expected to be required. What happens when we relax this constraint and allow genomes to change? What lessons can we learn from this? In 1970, a classical set of experiments performed by Sol Spiegelman provided a unique instance of evolution toward simpler genomes and illustrated some key ideas on viruses, replicators, and the power of selection (Spiegelman 1970).

Using the $Q\beta$ phage, Spiegelman created the conditions for replication of RNA viruses in a cell-free context (figure 2.7). Specifically, RNA chains of this phage, with a genome of

---

[5]In reality, the situation is much more complex, since viruses also make use of cellular proteins, named chaperones that dynamically help in folding other proteins.

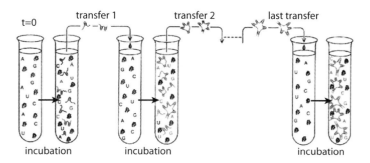

Figure 2.7. Evolution of shorter RNA genomes was observed in transfer experiments using cell-free conditions where shorter $Q\beta$ molecules generated by mutation won a fitness advantage by the virtue of being shorter and replicating faster.

$\nu \sim 4{,}500$ nucleotides length, were placed in a test tube along with free nucleotides and the $Q\beta$ replicase required to replicate the RNA chains. In these conditions RNA molecules can make copies of themselves, along with mutations caused by the error-prone replication machinery. Among other types of mistakes, shorter sequences can be generated. After one of these incubation processes has been completed and no more RNA chains are formed, a sample of the final soup is transferred to a new test tube under the same original conditions.

These transfers are repeated a number of times. As this artificial selection experiment proceeded, it was found that shorter and shorter sequences were obtained after each round of selection. Since no cells are present, any part of the $Q\beta$ genome not required to bind to the replicase becomes superfluous and at the end of the experiment the resulting chains had just $\nu \sim 200$ nucleotides and were completely unable to infect cells.

The experiment is relevant for several reasons. On the one hand it fully illustrates the power of selection, leading to the survival of the fastest: the shorter, the better. No other trait than fast speed (and thus short length) is at work. This actually corresponds to the simplest standard replicator competition model, which can be described as a set of n equations,

$$\frac{dx_i}{dt} = f_i x_i - x_i \Phi(t), \qquad (2.9)$$

where $x = (x_1, \ldots, x_n)$ indicates the populations of replicators (such as different RNA sequences in Spiegelman's experiment), each one growing at a given rate f_i. Here no mutations are considered and only competition is introduced in the last term, $\Phi(x)$, of the right-hand side. If we impose the condition $\sum_i x_i = 1$, which provides a population limit, it is easy to see that the sum over all the populations gives

$$\sum_{i=1}^{n} \frac{dx_i}{dt} = \frac{d}{dt}\left(\sum_i x_i\right) = \sum_{i=1}^{n} f_i x_i - \sum_{i=1}^{n} x_i \Phi(t), \qquad (2.10)$$

and, since $\sum_i x_i$ is constant, this gives

$$\Phi(t) = \sum_{i=1}^{n} f_i x_i, \qquad (2.11)$$

which is nothing but the average replication rate $\langle f \rangle$, and thus the previous set of equations can be written as

$$\frac{dx_i}{dt} = (f_i - \langle f \rangle) x_i, \qquad (2.12)$$

which has one single winner: the one with the highest replication rate. Since shorter strings are expected to be faster replicators, the expected outcome of this simple model is that smaller genomes must be the winners.

What kinds of viruses are selected, beyond those with small size? Further experiments (Mills et al. 1967) revealed that as RNA particles became shorter, they also were less infective, and at some point most sequences were simply unable to infect their original host cells. In other words, once freed from the host cell context, all elements required for the virus life cycle associated to infection are simply removed as useless. No cell context, no infection.

If we return to von Neumann's scheme of a formal replicating machine, Spiegelman's experiment has an interesting implication. Once the virus becomes independent of all host-dependent components, the context-free machinery gets closer to the "virus as software": a tape being read by a molecular machine that makes new copies. What is the lower bound? One key requirement for replication is the molecular recognition between RNA and its replicase. A too-short chain would fail to be recognized and thus no replication would take place (Adami 2006).

So far, the biology of cells and viruses seems to be about replicators and their parasites. Cells replicate by means of a regulated machinery that follows von Neumann's scheme. Viruses instead would be the outcome of the rich molecular toolkit provided by the cellular context. The picture, however, is more interesting, and the separation between viral and cellular life less sharp than it seems.

2.5 The Space of Replicators

The result of von Neumann's inquiry was a logical scheme that captures what is really crucial for an autonomous entity to complete a replication cycle. It is probably not an accident that here theory and molecular biology match quite well: perhaps the logic of self-replicating life has only one logical form. We use this theoretical picture because we would like to understand this fundamental property of living matter as a key for answering the question "what is life?" And since any answer will need to provide the boundaries of the problem, we can say that viruses define the coastline of living complexity. It could be argued that to answer our previous question we just need to answer first another question, namely: "are viruses alive?" But it turns out that things are not easy here either.

Viruses, as we have already discussed, can replicate themselves provided that a given cellular context is present. However, the broad spectrum of interactions among viruses and their hosts and an accurate list of properties defining life reveal a much more tangled picture (Koonin and Starokadomsky 2016). As an example, viruses are often considered nonliving because they lack the metabolic activity that is required to maintain cellular structures. On the other hand, living cells also include potential dormant states that are practically considered alive but are no less inert than viral particles outside their hosts. In these dormant systems, no growth or detectable metabolism will take place over very long periods of time. Are these dormant phases alive?

To further explore this question, Koonin and Starokadomskyy have used the *replicon paradigm* to properly define the space of possible forms of living organization where the degree of autonomy and selfishness in particular can be used as qualitative axes (Koonin and Starokadomskyy 2016). The starting point requires defining (and distinguishing between) replicons and replicators. The latter can be identified with autonomous replicating units (Dawkins 1982; Szathmáry 2006) displaying some autonomy with respect to genome replication. To a large extent, this corresponds to von Neumann's formulation, since both definitions assume that the "genome" is associated—in a stable manner—with a replicator. As defined, a replicator can contain a genome but also a number of additional "replicons", i.e., units of replication lacking the autonomy of a full replicator. As an example, an eukaryotic cell is a replicator carrying several replicons, whereas the genome of an intracellular organelle is a true replicon. Clearly a major feature that needs to be considered here is the degree of autonomy displayed by the replicator. One could easily conclude that viruses are nonautonomous replicators whereas cells are equipped with full autonomy. But a close analysis questions this view.

The previous definition suggests an almost binary split between the two types of units; a careful consideration of the properties of different replicators gives a richer and less clear-cut picture. Following again Koonin and Starokadomskyy (2016) we define a three-dimensional space of replicators (figure 2.8a) where

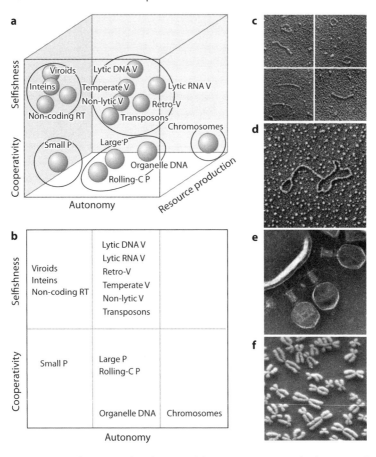

Figure 2.8. The space of replicators. (a) Here we arrange the location of different classes of replicating systems, from the smallest to the largest, using three basic coordinates involving the degree of selfishness (vertical axis) as well as autonomy and resource production. The projections reveal discrete classes. The diagram is inspired in Koonin and Starokadomskyy (2016). In (c–f) we display four examples of replicons (see text for details). *P* corresponds to DNA plasmids while *V* corresponds to viruses.

three axes are used to characterize (in an approximate way) the locations of different replicators. The vertical axis introduces the degree of selfishness: lower values are associated to cooperative replicators; more parasitic replicators occupy the upper part. A second axis calibrates the degree of autonomy displayed by the different replicators. Finally, a third axis introduces the number of resources required (or modified) to complete the replicators, reproduction. Despite the three axes being continuous, the actual locations of known replicators seem clustered in distinct subsets, as shown in figure 2.8b, where only the projected planes are shown. These four classes plus chromosomes are highlighted by means of light closed lines.

Looking at the autonomy-selfishness plane, we see that four classes are delineated by the criteria of presence or absence of signals for replication and/or transposition and the respective protein machinery. In this projection, five occupied quadrants are defined. The upper and lower domains would separate selfish from cooperative systems, whereas the three sets within the horizontal axis reveal three levels of autonomy, involving (from left to right) the types of signals enabling the replicative autonomy. Two basic classes can be defined either as parasitic or cooperative. Here the first class includes some obvious members such as lytic viruses that grow and eventually destroy their hosts and others (such as retrotransposons) that simply spread within their host genomes (sometimes constituting the majority of them). Plasmids,[6] on the one hand, are strongly dependent on the host replication cycle and can provide advantages to their hosts, thus acting as cooperators. Chromosomes, on the other hand, are the cellular replicators encoding a whole repertoire beyond the pure replication function and are clearly separated in the autonomy axis.

[6]A plasmid is a small DNA molecule that lives inside a cell and is physically separated from chromosomal DNAs and can replicate independently. It is most commonly found in bacteria as a small circular, double-stranded DNA molecule. Plasmids also exist in archaea and eukaryotic organisms. In nature, plasmids often carry genes, for example, antibiotic resistances or virulence factors, that may benefit the survival of the host organism (Thomas and Summers 2008).

The third axis is a useful, orthogonal dimension that also defines a potential continuum. Here we introduce the number of resources that are produced and/or modified by the corresponding replicator. Here two major categories are present (Koonin and Starokadomskyy 2016). One group includes complex replicators that make (or import) either all or part of the resources, as one would expect from cell-like entities. The second broad class encapsulates those simple replicators lacking any of these features, which would correspond to virus-like, parasitic entities. In this context, our third axis weights the resource autonomy of different replicators.

But here there are blurred boundaries that change everything. Many life forms clearly fitting the cell-like class require a cellular or multicellular host from which energy and resources are extracted. One example of such blurred limits is represented by endosymbiotic bacteria that live inside the cytoplasm of their host cells and are absolutely dependent on the host cell for replication and survival (Archibald 2015). An alternative example at the other extreme is represented by the cyanophages, a particular class of large bacteriophages that infect marine cyanobacteria of the genera *Synechococcus* and *Prochlorococcus* and that contain in their genomes key genes involved in the light harvesting apparatus used by these bacteria and plant chloroplasts (Clokie and Mann 2006). But the most important outsider in this scheme is the "nonproducers" that carry very large genomes coding (among other things) for enzymes and other key molecules that affect cellular metabolism and biosynthetic pathways. These large-sized nonproducers, particularly members of the weird *Mimiviridae* family, challange the binary scheme by placing themselves in an intermediate position.

In 1992 microbiologists from the Universities of Leeds and Marseille described what they thought was a tiny gram positive bacteria infecting the amoeba *Acanthamoeba polyphaga*, and named it *Bradfordcoccus*. A decade later, Claverie and co-workers

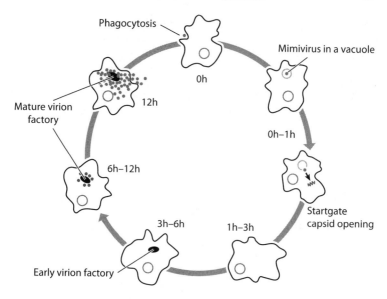

Figure 2.9. Life cycle of a giant virus. Here the sequence for the mimivirus *Acanthamoeba castellanii* is shown. This 12-hour cycle involves several marked phases that are illustrated by some snapshots obtained from electron microscopy. Redrawn from Claverie and Abergel (2010).

published an astonishing paper in *Science* (Raoult et al. 2004) concluding that the tiny bacteria in reality was a giant virus, as never seen before, which indeed belongs to Baltimore's group I (i.e., with a dsDNA genome). This virus was name *Mimivirus* and was the first of its class. This was the first in a run of discoveries in which larger and larger viruses, all sharing genomic properties and conserved genes, were found and characterized. The largest one added to the family was the huge *Pandoravirus*, with a genome size similar to that of some eukaryotic parasites (Philippe et al. 2013). As was the case for the *Mimivirus*, the *Pandoravirus* was also known before its characterization as a virus. In 2008 a German clinical microbiologist, Patrick Scheid, studying amoebae living in the contact lenses of a woman suffering from ocular keratitis described what he called endocytobionts, particles of about 1 μm size, living inside the amoebae. All these giant viruses infect unicellular amoeba. These giant viruses are so different from any other viruses that they share a characteristic property of cellular organisms: they have their own parasites.

Two classes of molecular parasites hosted by mimiviruses have been so far described. First, the Sputnik virophages, named analogously to the bacteriophages that infect bacteria (La Scola et al. 2008). Sputnik satellites have a dsDNA genome, larger than some bona fide viruses, of about 18 kb and encoding for about 20 proteins. Sputnik, however, parasitizes the functions encoded by its helper mimivirus and has been shown to be involved in the transfer of genes between mimivirus species. Second, as cells, the mimivirus genomes also carry transposable elements, in this case named transpovirons (Desnues et al. 2012). Transpovirons are linear DNA elements of about 7 kb that encode about seven protein-coding genes needed for their own mobilization. Some of these genes are homologous to Sputnik's own genes. All mimiviruses share some structural, genomic, and functional similarities with some long-time known members of Baltimore's group I viruses (see chapter 1), for example, poxviruses infecting vertebrates (including humans) and arthropods. Well-known examples are the smallpox (variola), vaccinia and cowpox viruses. All together, these viruses are known as the nucleocytoplasmic large DNA viruses or NCLDVs. NCLDVs share a core set of 47 genes that include enzymes involved in DNA replication and repair, transcription factors, and proteins that interfere with host cell proliferation (Colson et al. 2013). Such a level of complexity, closer to the cellular machinery than to the standard view of viral software, is puzzling. We will return to this in chapter 7. Let us just mention for now that the virophage concept challenges the simple picture of viral software using cellular hardware to replicate itself. In fact, the ways virophages act on the *Mimivirus* molecular replication machinery strongly reminds us of the diagram of virus infecting cells.

2.6 Adaptation at High Mutation Rates

Although most viruses share striking similarities with computer programs, they depart from this analogy in several ways. On the one hand, the parasitic software carried by these tiny agents makes no sense unless we place it in the context of a preexisting cell machinery that is hijacked to guarantee virus replication. As with the rest of biology, viral life cycles make sense only under the light of evolution. And evolution requires both variation and selection, even when dealing with man-made computer viruses (see chapter 8). Mutation is not a component of software. On the other hand, almost by definition, programs need to preserve even their minor components from mistakes. Viruses instead might need to face the challenge of a host organism that fights back, and as a consequence an arms race between viruses and their hosts ensues (see chapter 4). One particularly remarkable outcome of such an arms race is that instability is pushed to the limit.

In previous sections of this chapter, we mentioned the prominent role played by information and computation as defining traits of living entities. Because of their intrinsic simplicity, and due to their strong dependence upon a host molecular machinery in order to replicate, viruses are unique dynamical systems. One particularly important trait of RNA viruses is their high mutation rates, much higher than any other rates exhibited by cellular systems and a consequence of the lack of repair mechanisms associated to their RNA-dependent RNA polymerases. This enzyme catalyzes the replication of RNA from an RNA template, and mutation rates per nucleotide and replication cycle are in the range 10^{-4}–10^{-5} (Sanjuán et al. 2010). In DNA-based systems, such as cells, the process of DNA polymerization is usually associated to a proofreading and repair mechanism that effectively reduces mutation rates to a range 10^{-8}–10^{-11}, ensuring a controlled replication cycle (Drake et al. 1998). In stark contrast, RNA polymerases lead to high mutation rates that are orders of

magnitude larger. Since high mutation carries a burden of phenotypic errors, that implies that many resulting viral genomes can
contain deleterious changes leading to nonviable viral particles.

Mutation is a crucial component of evolution, as genetic
variability is the fuel on which natural selection operates to adapt
populations to their environment. In this sense, an error-prone
polymerase can be seen as useful in order to keep pace with the
always changing environmental conditions in which RNA viruses
live (Domingo 2000). However, keeping in mind that mutation
is a random process independent of the value that the mutations
generated may have in the future generations, mutation itself
is a double-edged sword: too many mutations per genome may
simply drive fitness levels so low that they are not compatible
anymore with a succesful replication. Therefore, mutation rates,
like any other trait, themselves have evolved and have been
optimized for the lifestyle of RNA viruses: just high enough
but not more (Elena and Sanjuán 2005). For RNA viruses, a
heterogeneous population results in a so-called *viral quasispecies*.
A viral quasispecies can be seen as a swarm of genomes dominated
by a *master sequence* that may coincide with the average sequence
of the population, the consensus sequence.

The quasispecies structure has many implications on the
biology of RNA viruses. The most important is that the swarm
stores a reservoir of phenotypes crucial for coping with environmental uncertainties: within the context of the virus infection and
pathogenesis, that includes the host responses tied to immunity
but also others such as tissue specificity or resistance to drugs
(Andino and Domingo 2015; Domingo et al. 2012; Lauring and
Andino 2010; Holmes 2010).

One particularly unexpected consequence of the quasispecies
nature of viral populations is deeply tied to information. This
is known as the *error catastrophe* problem (Eigen 1971; Eigen
et al. 1987; Schuster 1994; Domingo and Holland 1994) and
is deeply connected with phase transitions in physics. It was

originally defined within the context of an abstract population of mutating and replicating molecules (or entities), generically called *replicators*, competing for limiting resources. More precisely, Eigen and Schuster considered a (large) population of strings (genomes or polymers) where each sequence can replicate at some rate. Replication rate will be sequence-dependent and the relation between sequence and growth rate should be expected to be complex (see chapter 3). Additionally, we assume that every time a chain replicates, mutations can occur at a given rate μ. Eigen (1971) predicted that mutation imposes a limit on the amount of information (in terms of genome length, ν) that is consistent with stable information. Specifically, it was shown that there is a critical mutation rate $\mu_c \sim 1/\nu$ beyond which no Darwinian evolution can occur. His theoretical work (see below) thus made a key prediction: no viable sequences would be observable for mutations higher than the critical one, i.e., for $\mu > \mu_c$. In that case, random drift would be observed. Instead, below the threshold, selection operates and information can be maintained in stable ways. Available information confirms this inverse relationship, as shown in figure 2.10a. Mutation rates decrease as an inverse power law of genome length. RNA viruses exhibit the highest rates, orders of magnitude larger than DNA viruses, as dramatically illustrated by viroids. These are extremely small viruses, equipped with a minimal, non-protein coding genome (figure 2.10b) that infects plants and uses their RNA polymerases to replicate. As larger genomes are analyzed, mutation rates rapidly decay. Two questions emerge: What is the origin of such a relationship? What are the limits (if any) on mutation rates in viruses?

2.7 Viral Quasispecies

Because of their high mutation rates, RNA virus populations are highly heterogeneous, thus defining a cloud of mutationally

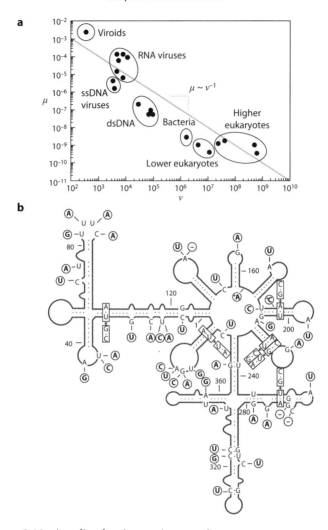

Figure 2.10. A scaling law in per-site mutation rate versus genome size. (a) Includes chosen examples of RNA viruses, including the *Chrysanthemum chlorotic mottle viroid* (CChMoVd), with a 399-nucleotide genome, with a secondary structure displayed in (b). Larger genomes are represented by both ss and ds RNA and DNA viruses, microbes, and a few eukaryotes (adapted from Gago et al. (2009)). The continuous line is used to highlight the inverse law linking mutation rate μ and genome length ν predicted by the error threshold theory.

related sequences (Eigen et al. 1989; Domingo et al. 1989, 2012). There is thus a mutant spectrum, rather than a genome or a small collection of genomes, and it is more than simple sets of independent mutant sequences. Instead, viral quasispecies involve a repertoire of variants that can help overcome selection pressures. A mathematical approach to the behavior of these mutant clouds was provided by the Eigen-Schuster model. This model considers a set of populations $\{x_i\}$ representing the abundance of different genomes, changing in time by the following set of equations:

$$\frac{dx_i}{dt} = \sum_j f_j \mu(j \to i) x_j - \Phi(\mathbf{x}, t) x_i, \qquad (2.13)$$

where x_i indicates the fraction of the population associated to the ith mutant genome equipped with an M-letter alphabet (here $i = 1, ..., n$, where $n = M^v$ is very large) so that a normalization condition applies, namely, $\sum_j x_j = 1$. Here f_j is the growth rate of the jth mutant, $\mu(j \to i)$ is the probability of having a mutation from sequence j to sequence i, and $\Phi(\mathbf{x})$ is the average fitness associated to the population vector $\mathbf{x} = (x_1, ..., x_n)$, i.e.,

$$\Phi(\mathbf{x}, t) = \sum_j f_j x_j \left(\sum_j \mu(j \to i) \right) = \sum_j f_j x_j = \langle f \rangle.$$

$$(2.14)$$

This model can sometimes be treated analytically under a well-defined set of conditions, showing that the population structure corresponds to a cloud of sequences (Eigen 1971; Eigen et al. 1987; Schuster 1994). Let us consider a specific case that will illustrate how mutation can sharply limit the length of genomes and thus the amount of information stored in a quasispecies.

Consider the general model described above and now use only two values for the replication (fitness) of genomes, namely f_m for the master and f for any other sequence (i.e., $f_1 = f_2 = ... = f_n = f$), where n is very large ($n \gg 1$). Hereafter we assume

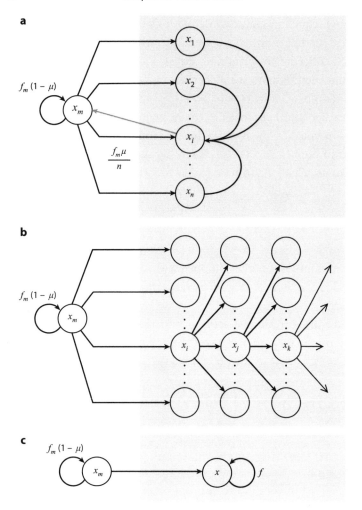

Figure 2.11. The quasispecies model involves a system of mutationally connected genomes whose populations are here indicated as x_k. (a) Indicate by x_m the master sequence population, which makes copies of itself at a rate $f_1(1 - \mu)$ while mutating into n different possible sequences $x_1, \ldots, x_n$. Mutations from one strain are shown as arrows (only in one direction: another arrow should indicate back-mutation). In a more realistic setting system, x_m can mutate into a small number ν of sequences, while most of the M^ν strings are essentially unavailable. If we consider

that $f_m > f$, i.e., that the master sequence is more efficiently replicated than any other one. We will also asume for now that all mutation rates are the same among the nonmaster strings. Specifically, we have $\mu(i \to j) = \mu/n$ for $i \neq j$ and $\mu(i \to j) = 1 - \mu$ for $i = j$. In this case (described by figure 2.11a) we can split our system of equations into two sets: the master sequence and the rest. The system presented below is only a simplified approximation of the space connecting different genomes. A more accurate picture is provided by figure 2.11b: from a given population x_j, mutation will not lead back to the master sequence or will occur with difficulty (see chapter 3). In this case, we will have $x_m + x = 1$, where we use $x = \sum_j x_j$. For the master sequence we get

$$\frac{dx_m}{dt} = f_m(1 - \mu)x_m + \sum_{j=1}^{n} f_j \frac{\mu}{n} x_j - x_m \Phi(t), \qquad (2.15)$$

whereas the set of equations for the nonmaster sequences reads:

$$\frac{dx_i}{dt} = \frac{f_m \mu}{n} x_m + \sum_j f_j \frac{\mu}{n} x_j - x_i \Phi(t). \qquad (2.16)$$

It is easy to show that the first equation can be simplified to

$$\frac{dx_m}{dt} = f_m(1 - \mu)x_m + \frac{f\mu}{n}(1 - x_m) - x_m \Phi(t) \qquad (2.17)$$

$$\approx f_m(1 - \mu)x_m - x_m \Phi(t), \qquad (2.18)$$

where we have used $\mu/n \ll 1$. For the nonmaster system, we have

$$\frac{dx_i}{dt} = \frac{f_m \mu}{n} x_m + f(1 - \mu)x_i + \sum_{j \neq i} \frac{f\mu}{n} x_j - x_i \Phi(t), \qquad (2.19)$$

Figure 2.11. (*Continued*) that all nonmaster sequences replicate at the same rate f under the (b) picture, the system can be collapsed into a two-component graph (c), where x is the sum of all nonmaster populations.

and since our goal is to describe the dynamics of *all* these populations together, we take the sum over all genomes, i.e.,

$$\sum_i \frac{dx_i}{dt} = \sum_i \frac{f_m \mu}{n} x_m + \sum_i f(1-\mu)x_i \quad (2.20)$$

$$+ \sum_i \frac{f\mu}{n} \left(\sum_{j \neq i} x_j \right) - \sum_i x_i \Phi(t) \quad (2.21)$$

$$= \mu f_m x_m + f(1-\mu)x + f\mu(x - x_i) - x\Phi(t); \quad (2.22)$$

given the homogeneous mutation rates, a stationary distribution will lead to $x_i \approx x/n$, and thus we can approximate $x - x_i \approx x$, which leads to

$$\frac{dx}{dt} \approx \mu f_m x_m + fx - x\Phi(t) \quad (2.23)$$

and, as a result of these approximations, a two-compartment model (figure 2.11c) that can be collapsed into a one-dimensional system (Swetina and Schuster 1982):

$$\frac{dx_m}{dt} = f_m(1-\mu)x_m - x_m \Phi(x_m, x) \quad (2.24)$$

$$\frac{dx}{dt} = f_m \mu x_m + fx - x\Phi(x_m, x). \quad (2.25)$$

As we did above for the replicator model, the condition $x_m + x = 1$ leads to $\Phi(t) = \langle f \rangle = f_m x_m + fx$. Using this function, and since $x = 1 - x_m$, we obtain, after some algebra, an equation for the master sequence:

$$\frac{dx_m}{dt} = x_m \left[(1 - x_m)(f_m - f) - \mu f_m \right]. \quad (2.26)$$

Two alternative equilibria are possible (from $dx_m/dt = 0$), namely $x_m = 0$, associated to an extinct master sequence, and a nontrivial state

$$x_m = 1 - \frac{\mu f_m}{f_m - f}. \quad (2.27)$$

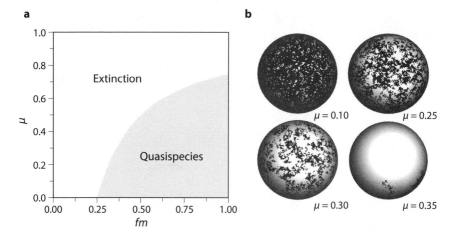

Figure 2.12. The error catastrophe phase transition, illustrated by its simplest version (see text). In (a) we plot the two phases associated to this model, using mutation rate μ and f_m as key parameters. These phases are separated by a critical line $\mu_c(f_m, f)$. A spatial, discrete model involving two states (black and white, indicating master and nonmaster sites) provides a complementary picture of the transition.

The latter solution will be positive (and the master sequence will be present) provided that $x_m > 0$, and this will occur if the mutation rate is lower than a critical value:

$$\mu < \mu_c = 1 - \frac{f}{f_m}. \qquad (2.28)$$

This expression defines the boundary separating two well defined phases, shown in figure 2.12a. Within the quasispecies phase (gray area) master genomes will be present, along with a tail of mutants. Once the boundary is crossed, all master sequences disappear (despite their higher replicating efficiency).

This model can be mapped into a discrete lattice model where mutational and replication events can be seen as reactions. Consider an $L \times L$ two-dimensional grid Ω. The states $S(i, j)$ of a site $(i, j) \in \Omega$ will be indicated by M (master string) and N

(nonmaster). Following the mean field model described above, it is easy to see that the terms associated to the previous equations are described by the reactions

$$M + N \xrightarrow{P_m(1-\mu)} M + N \tag{2.29}$$

$$M + N \xrightarrow{P_m\mu} N + N \tag{2.30}$$

$$N + M \xrightarrow{P} N + N, \tag{2.31}$$

where the probabilities P_m and P replace the previous rates of replication f_m and f, respectively.

The model exhibits the same kind of transition (although there is some spatial dependency that leads to corrections), but also allows us to visualize the complexity of the fluctuations exhibited by the master equation as the critical boundary is approached. In figure 2.12b we show several examples of the spatial dynamics of our cellular automaton for different μ values, after fixing $P_m = 1$ (i.e., if chosen, the master always replicates) and $P = 0.5$. As we increase mutation rates, the number of master sites is reduced, but the domains of master sites also experience complex fluctuations. Complex clusters of many different sizes can be observed. These fluctuations are nothing but those known from the physics of phase transitions (Goldenfeld 1992; Solé 2011). Actually, it has been shown that the quasispecies formalism can be expressed in terms of one of the classic models of phase transitions, the so-called Ising model (Leuthäusser 1986, 1987). The Ising model was successfully used to predict the behavior of ferromagnetic transitions when temperature T is increased beyond a critical threshold T_c. Once $T > T_c$, magnetization is lost, and critical fluctuations are observed close to T_c as a consequence of a conflict between order from interactions and thermal disorder. In our context, mutations are also a source of disorder, while accurate replication is responsible for order (a stable master population). Here too we find a surprising connection between the dynamics of viruses and properties of matter and their phases.

2.8 Critical Genome Size

As discussed above, a scaling law seems to connect the mutation rate of a given genome and its mutation rate (figure 2.10a). We are now in a position to mathematically derive this remarkable dependence. In our previous calculations, μ defines the mutation rate (for each replication round) per genome. But the accurate definition of mutation requires considering the units forming each string of length ν. If μ_b is the mutation rate per site, it is not difficult to see that it is related to μ by

$$\mu = 1 - (1 - \mu_b)^\nu. \tag{2.32}$$

Here $p = (1 - \mu_b)^\nu$ is the probability that none of the units are mutated; the difference $1 - p$ is the probability that some unit (and thus the genome) does mutate. Since μ_b is typically small, we can use the approximation[7]

$$\mu \approx 1 - e^{-\mu_b \nu} \approx 1 - \mu_b \nu. \tag{2.33}$$

If we return to the previous critical condition for mutation rates and write it down as a function of μ_b, we have

$$\mu_b^c = \frac{\alpha}{\nu}, \tag{2.34}$$

where $\alpha > 0$ is a constant. The last expression actually corresponds to the observed inverse decay of mutation rates as an inverse of their genome size (figure 2.10a). In particular, RNA viruses have been shown to follow this inverse law, thus being consistent with Eigen's theory. Many different extensions of these models have been proposed (Manrubia et al. 2010) considering, for example, more complex landscapes (Saakian et al. 2006; Saakian and Hu 2006; Schuster 2016), spatial dynamics (Altmeyer and McCaskill 2001; Pastor-Satorras and Solé 2001;

---

[7]Here we made use of the first two terms of the Taylor expansion of the exponential function, i.e., $e^x \approx 1 + x + x^2/2! + \ldots$.

Aguirre and Manrubia 2008), or the role played by secondary structure on phenotypic thresholds (Ancel and Fontana 2000; Stitch et al. 2007), to cite just two.

We can appreciate that the model is very simple and yet very rich in dynamical complexity and—what is more important— makes a well-defined prediction: There is an upper bound to the rate of disorder that an adaptive system can tolerate. Nowadays we know that RNA viruses live on the edge of catastrophe (Schuster 1994; Domingo and Holland 1994; Solé et al. 1996), taking advantage of the mixture of order and disorder characteristic of critical points. An interesting consequence of this result is that, since the error threshold is a phase transition point, it defines a sharp boundary separating two phases, one of which involves the loss of the master sequence. Once this occurs, a uniform diffusion of populations will take place within the non-master set, losing the quasispecies structure, and no Darwinian adaptation can occur. The theory thus predicts that even a small increase of mutation rates beyond μ_c leads to the extinction phase. Experimental tests using mutagenic drugs fare consistently with this prediction, indicating that a potential path (particularly combined with other therapies) to fight viral populations could increase mutagenesis (Loeb et al. 2000; Perales et al. 2012). This provides an elegant example of how phase transitions could be exploited to fight disease in nonstandard ways.[8]

---

[8]More generally, the dynamical behavior of RNA quasispecies is relevant to several aspects of therapeutics (Perales et al. 2012), including the outcome of multidrug treatments and the role played by population fluctuations.

3

LANDSCAPES

3.1 Climbing High

As we already discussed in chapter 1, RNA viruses offer a unique opportunity for exploring long-term evolutionary dynamics under experimentally controlled conditions, owing to their intrinsic simplicity (Domingo and Holland, 1994). They are successful, to a large extent, because they are simple, small-sized, and fast-replicating, but especially because of their enormous plasticity and adaptability to changing environments. Such plasticity stems from the high mutation rates during RNA genome synthesis and rapid replication. In order to describe and model their dynamics, there are several levels of approximation that can be used, from the molecular shapes of their genomes to coarse grained descriptions where just a few parameters (such as mutation and replication rates) encapsulate the minimal phenotypic traits of each strain.

Using viruses as model systems for tackling fundamental questions in evolutionary biology (and complex systems) may sacrifice some realism for the benefit of a clear understanding of key mechanisms. In particular, the complex structure of the virus' environment (its host organism) and the fact that viral particles are not just strings of nucleotides are often neglected. Moreover

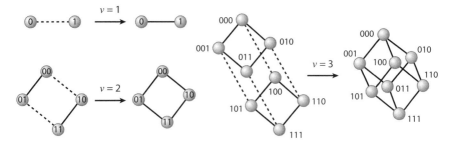

Figure 3.1. Sequence spaces and their iterative construction in a two-letter alphabet system. The iterative generation of one landscape, $\Sigma^{\nu+1}$, from the previous Σ^{ν} (with strings connected through solid lines) can be interpreted geometrically as a translation (dashed lines) where an additional bit is added to each previous string to create a new landscape with an additional dimension.

(see next section), the details of the virus-host interactions and the molecular peculiarities of the viral life cycle are not taken into account. And yet, even under very simplistic assumptions, relevant properties of virus dynamics can be captured.

The evolution of a virus population with a genome of length ν can be visualized as a dynamical process of growth, competition, and selection taking place in the sequence hypercube $\mathcal{H}^{\nu}$ defined as the set of ν-length strings:

$$\mathcal{H}^{\nu} = \{(S_1, S_2, \ldots, S_{\nu}) \mid S_k \in \Sigma\}, \tag{3.1}$$

where Σ is the alphabet of the genetic code, which can be the RNA one, $\Sigma = \{A, G, U, C\}$ or, in a simplified approach used in modeling, a binary (Boolean) set $\Sigma = \{0, 1\}$. As defined, we have a landscape given by the Cartesian product

$$\mathcal{H}^{\nu} = \Sigma \times \Sigma \times \ldots \times \Sigma = \Sigma^{\nu}. \tag{3.2}$$

For a binary landscape, it has a total of $|\mathcal{H}^{\nu}| = \Sigma^{\nu}$ elements. The iterative procedure to generate these sequence spaces is summarized in figure 3.1. Here $\mathcal{H}^{\nu}$ is generated from $\mathcal{H}^{\nu-1}$

by means of an expansion of the latter that can be interpreted geometrically as a translation where an additional bit is added to each previous string (thus the previous number of genotypes gets multiplied by 2). The situation is somewhat more complex when dealing with a 4-letter alphabet (bottom sequence), where each node in the higher-dimensional landscapes is made by a tetrahedron (Schuster 2009).

If we consider a population of strings $\{S_i\}$, $i = 1, \ldots, N$, each string $S_i = (S_{i1}, \ldots, S_{iv})$ of length v describes a digital genome, i.e., a sequence of purines and pyrimidines that incorporate only the linear information encoded by the string. In order to introduce further information concerning the functional relevance of this sequence, we need to introduce an additional mapping, namely the sequence-fitness measure defined as

$$f : \Sigma^v \longrightarrow U \subset \mathcal{R}, \tag{3.3}$$

which maps the string into a scalar value, i.e.,

$$S_i \in \Sigma^v \longrightarrow f(S_i). \tag{3.4}$$

The nature of this mapping is the central problem considered here. If this functional relation is such that bits are essentially independent from each other, no epistatic components will be at play. Instead, if the function f is such that different parts of the system influence others in some nontrivial ways, then epistasis[1] will be at play. Although the sequence space is comprised of an astronomically large, multidimensional set of points, we can start using a simple three-dimensional surface to illustrate the ideas developed below (figure 3.1).

The fitness landscape, an idea first introduced by evolutionary geneticist Sewall Wright and later extended by several other

---

[1]In evolutionary genetics, we talk about epistasis when the observed phenotype cannot be directly predicted from the contribution of each of its genotypic components. In other words, when the function linking genotype and phenotype is nonlinear. This will become clear in section 3.3.

authors (e.g., Kauffman 1990; Perelson and Kauffman 1991, and references therein), is defined in terms of some particular traits that are implicit in the virus particle phenotype and are usually described in terms of replication rate or infectivity. The landscape appears in most textbook plots as a multipeaked surface. Local maxima represent optimal fitness values, which can be reached through mutation from a subset of lower-fitness neighbors. Given an initial condition defined by a quasispecies distribution localized somewhere in the sequence space, the population will evolve by exploring nearest positions through mutation. In figure 3.2 three examples of idealized fitness landscapes are depicted, using real mountain peaks (left column) and a small-sized, four-bit hypercube (right column). They involve three prototypical cases. The first is the simplest, ideal scenario where there is just one global maximum as with Mount Fuji in Japan. If we consider the Boolean counterpart, the maximum is located at the $(1, 1, 1, 1)$ node and surrounded by lower-level fitness nodes that have smaller fitness values as we move from the peak. One simple way of getting this single-peak system is by using a fitness function to be defined by each individual string

$$S_i = (S_i^1, \ldots, S_i^v)$$

in the population, and is given by the linear function

$$f(S_i) = f_0 + \sum_{k=1}^{v} S_i^k, \qquad (3.5)$$

i.e., by adding the number of 1's of the given sequence. Here f_0 gives a minimal, ground level of fitness associated to the all-zeros case. How would we write this fitness function for the sharp peak landscape defined in the previous chapter? If S_m is the master sequence, with fitness f_0, and S_n any other string, having fitness f_1, it is not difficult to see that $f(S_i)$ can be written as: $f(S_i) = f_0 \Theta + f_1(1 - \Theta)$, where $\Theta = \prod_{k=1}^{v} S_i^k S_m^k$.

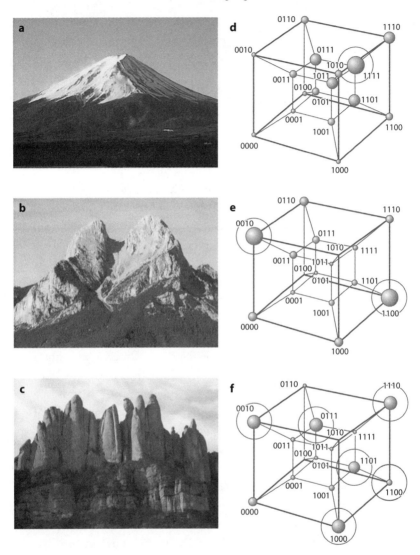

Figure 3.2. Fitness landscapes. In its classical formulation, fitness land-scapes are depicted as mountains with a given number of peaks in an abstract, two-dimensional landscape. Three examples from real land-scapes are shown: (a) Mt. Fuji, in Japan, (b) Pedraforca, and (c) Montserrat (both in Catalonia), illustrating fitness surfaces having one, two, and many peaks, respectively. In (d-f) the corresponding landscapes on a Boolean hypercube are depicted.

In these and other simple cases, the fitness function can be explicitly defined in mathematical terms, but in general it is more common is to simply assign a set of numbers connecting sequences and the adaptive value of their corresponding phenotypes. A more complex situation is shown in figure 3.2b, with a two-peak landscape where now two local optima are present. This picture has also been widely used to illustrate the idea of potential peaks that can be separated by a low-level fitness gap (the pass between peaks) and thus might be unreachable by mutation processes alone. A possible equivalence of this scenario on our Boolean cube is displayed in figure 3.2e. And this is just the simplest case for the large class of so called rugged landscapes characterized by the presence of many peaks, which would correspond to the third example shown in figure 3.2c, displaying the mountain range of Montserrat, in Barcelona. In practical terms, the simplest way of obtaining such a *rugged* landscape with multiple peaks is to assign a random number $\xi \in (0, 1)$ to each string. Many local optima will be generated[2] and thus many possible end points for the dynamics exist and thus a multiplicity of potential fitness peaks will be present (figure 3.2f). The nature and consequences of the ruggedness of a landscape are, as we will see, a central problem for our understanding of virus dynamics.

The third component of our theoretical approach to evolution on fitness landscapes requires us to define the dynamics of populations in sequence space. One approach is given by so-called *adaptive walks* (Kauffman and Levin, 1987). Consider a population Π of N strings. Once again a fitness function is introduced, $f(S_i) = f(S_{i1}, \ldots, S_{i\nu})$, but we now impose the

---

[2]An interesting feature of this model is that, because of its simplicity, it allows predicting some expected patterns of evolutionary dynamics (Kauffman 1990). If the fitness values are random and uncorrelated, i.e., if $f(S) = \eta \in [0, 1]$ is a random number with uniform distribution, it can be shown that the number of local optima (fitness peaks) is $M(\nu) = 2^\nu/(\nu + 1)$.

restriction that changes (mutations) occur by means of single, one-bit steps. For each generation in the algorithm, each string $S_i \in \Pi$ is chosen, along with one randomly chosen neighbor $S_j \in \Pi$ at a one-mutation distance. This is defined through the Hamming distance:

$$d_H(S, S') = \frac{1}{2} \left(v - \sum_{k=1}^{v} S_k S'_k \right). \qquad (3.6)$$

The fitness of each digital genome is computed and compared: if $f(S_i) > f(S_j)$, a replacement $S_i \to S_j$ occurs, given that such change would increase the fitness. The situation $f(S_i) = f(S_j)$ corresponds to a *neutral* walk: no fitness change will take place and thus the shift $S_i \to S_j$ can occur or not with the same probability. In this way, a random walk among sequences can take place and (unexpectedly) this will have important consequences on the evolution of viruses and their robustness (see section 3.5). A direct consequence of the adaptive walk dynamics is that populations can climb toward a local peak, staying there afterward.

More generally, an evolutionary dynamical system can be represented by a set of equations describing $x(S, t)$, namely the fraction of strings having a given sequence $S \in \Sigma^v$. It is not difficult to see that a general model[3] is given by a set of Σ^v differential equations:

$$\frac{dx(S, t)}{dt} = \sum_{S'} \mu(S' \to S) f(S') x(S', t) - x(S, t) \Phi(t). \qquad (3.7)$$

The terms of the equations are easy to identify: $\mu(S' \to S)$ is the probability of mutation from S' to S. Given the constant

---

[3] Here we just define the deterministic set of equations, but the problem can be easily generalized in several ways to include stochastic effects, which can play an important role.

population condition $\sum_s x(s, t) = 1$ the function Φ reads (see previous chapter):

$$\Phi(t) = \sum_{S'} f(S')x(S', t) = \langle f \rangle. \tag{3.8}$$

Despite the popularity of the fitness landscape picture interpreted as a more or less irregular surface, this view is, in most cases, misleading. Instead of this picture in which populations crawl across hills and valleys in search of higher locations, the right picture deals with high-dimensional sequence spaces formed by vast neutral networks linking equal-fit genotypes (Aguirre et al. 2011 and references therein). Among other things, the view of populations getting trapped on some local fitness peak vanishes. As we will show in this chapter, the structure of the fitness landscape is more interesting and can often lead to unexpected properties.

3.2 Symmetric Competition

Before we move into some intricacies associated to rugged landscapes, let us consider a simple experiment with RNA viruses. The example illustrates how the landscape picture can be helpful in explaining some counterintuitive observations (Clarke et al. 1994). In the experiment, two strains of the *Vesicular stomatitis virus* (VSV, figure 3.3a) were mixed and allowed to evolve and compete with each other. Specifically, two clonal populations of VSV having the same fitness were shown to exhibit steady gains in fitness using standard virus plaque assays (Novella et al. 1995).[4] As they evolved through time, each strain evolved by increasing

---

[4]Genetically marked monoclonal antibody resistant (MARM) clones of equal fitness to the wild-type VSV were used and their relative frequencies were monitored along passages by exposing the mixture to the corresponding antibody and counting the number of viruses capable of growing in its presence and absence, respectively.

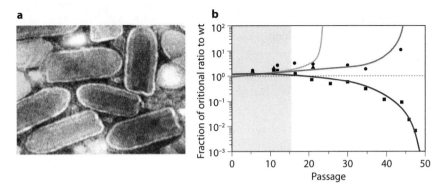

Figure 3.3. Experimental evolution of two competing clones of (a) VSV. In (b) we display three examples of the competition passages using BHK-21 cells as hosts. Here the points indicate the measured fractions of the two clones. Redrawn after Clarke et al. (1994).

its replication rate by an equivalent amount, indicating that both populations were able to keep competing while constantly changing. In figure 3.3b three outcomes (among many) of the experiment are shown. The general pattern found here is that both populations remain close until, after a given number of passages, their trajectories diverge.

The fact that both populations remain close despite both populations experiencing fitness increases is a counterintuitive result. Since these populations compete for resources, one would expect rapid divergences right from the beginning: the faster replicator should overcome the slower one. Such a change, where both populations display an increase in order to just keep in place, indicates that this is a good illustration of the famous Red Queen[5] effect (Van Valen 1973; Stenseth and Maynard-Smith 1984). Both competing populations grew showing the

[5] Just like the Red Queen in Lewis Carroll's *Through the Looking Glass*, where Alice and the Red Queen run faster and faster just to remain in the same place, each species in the experiment is forced to keep changing just to remain in the competition.

same steady increases in fitness[6] but, at some point, one of the populations suddenly dominated, excluding the other one. The winner of this competition process was not always the same. Sometimes one of the populations dominated, sometimes the other one (figure 3.3a). But they never coexisted. Why?

This question can be answered by using a general scenario where two quasispecies compete. This can be easily formulated in mathematical terms as follows. Here two different populations x^1 and x^2 are considered, each one formed by a set of mutants. Specifically, we have $\{x_i^k\}$ (with $i = 1, \ldots, n$; $k = 1, 2$), which compete for a given set or resources, with $x^k = \sum_i x_i^k$. Assuming that mutations occur with the same rates for all strains, we can write down the following system of equations (see previous chapter):

$$\frac{dx_i^k}{dt} = f_i^k(1 - \mu)x_i^k + \sum_{j \neq i} f_j^k \mu(j \to i)x_j^k - x_i^k \Phi(t) \quad (3.9)$$

$$= f_i^k(1 - \mu)x_i^k + \frac{\mu}{n}(x_k - x_i^k) - x_i^k \Phi(t) \quad (3.10)$$

$$\approx x_i^k \left(f_i^k(1 - \mu) - \Phi(t) \right). \quad (3.11)$$

The last equation is the replicator equation already introduced in the previous chapter.

In the context of the experiments with VSV, we are considering two clonal populations competing for the cells in the cell

---

[6]In order to estimate these fitness changes, the population was sampled at the end of each passage and cultivated along with the wild-type. After their growing together, the fitness is estimated from the relative difference in populations of each type.

culture (specifically BHK-21 cells). In this case, summing all the equations for x_i^k we have $dx_k/dt = \sum_i (dx_i^k/dt)$, i.e.,

$$\frac{dx_k}{dt} = \sum_{i=1}^{n} x_i^k$$

$$\times \left(f_i^k(1-\mu) - \sum_{i=1}^{n} f_j^1(1-\mu)x_j^1 - \sum_{i=1}^{n} f_j^2(1-\mu)x_j^2 \right) \tag{3.12}$$

$$\approx \sum_{i=1}^{n} x_i^k \left(e_i^k - \sum_{i=1}^{n} e_j^1 x_j^1 - \sum_{i=1}^{n} e_j^2 x_j^2 \right), \tag{3.13}$$

where we use $e_i^k = f_i^k(1-\mu)$. By defining the average value

$$\langle e_k(t) \rangle = \frac{\sum_{j=1}^{n} e_j^k x_j^k}{\sum_{j=1}^{n} x_j^k} \tag{3.14}$$

after some algebra we obtain a set of equations:

$$\frac{dx_1}{dt} = \langle e_1 \rangle x_1 \left(1 - x_1 - \beta_{21} x_2 \right) \tag{3.15}$$

$$\frac{dx_2}{dt} = \langle e_2 \rangle x_2 \left(1 - x_2 - \beta_{12} x_1 \right), \tag{3.16}$$

where we define

$$\beta_{21} = \frac{\langle e_2 \rangle}{\langle e_1 \rangle} \quad \beta_{12} = \frac{\langle e_1 \rangle}{\langle e_2 \rangle}. \tag{3.17}$$

This set of equations is a special case of the well-known Lotka-Volterra model of species competition (Case 2000). This model is known to exhibit two main types of solutions: species either coexist or exclude each other. In particular, for the symmetric case

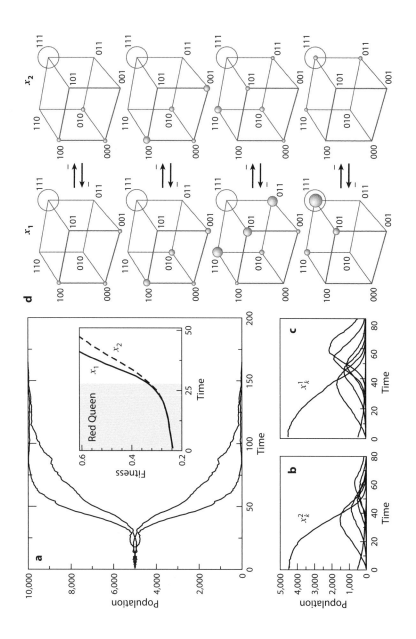

$\beta = \beta_{ij}$ it is easy to show that for $\beta < 1$ coexistence occurs, while for $\beta > 1$ only one of the species can win, with the other decaying into extinction. In our model, we can see that $\beta_{12} = 1/\beta_{21}$, and the analysis of this system reveals that coexistence is never possible, independently of the specific values of the competition coefficients (Solé et al. 1999). The previous experiments were also simulated (Solé et al. 1999) using a bit-string model that considered a population of N binary sequences. This model allows us to keep the mutation terms required to move through the fitness landscape. At each time generation (passage), the algorithm repeated N times the following rules: a random string, say S_i, of the population was chosen for replication. Replication is proportional to $f(S_i)$ and takes place by replacing another string, say S_j, by a copy of S_i. Errors occur at a rate μ (per bit and replication cycle).

The model consistently reproduced the experimental results (figure 3.4a-c). The populations were initialized with identical genomes and numbers of strings, in such a way that their initial fitness was low. During the simulations, strings were competing for the N available sites. For a while, both populations gained fitness (inset of figure 3.4a) but remained close in terms of population size (figure 3.4a) since replication rates were not strong enough. However, after a given number of passages, waves of growth and decay emerged (figure 4b-c) as fitter variants appeared by mutation. At some point (also predictable, as in the

Figure 3.4. Modeling the Red Queen dynamics of the VSV competition experiment with equal fitness strains. In (a–c) we display several examples of the time evolution of the bit-string model using digital genomes of length $\nu = 16$ and $\mu = 10^{-3}$. After some time, the two populations diverge (a) while diffusing over sequence space. In (b-c) we show the different population waves of more fit strings. At the beginning both strains show similar waves of replacement, but eventually one of them gets extinct. A simplified picture of the two coupled landscapes is depicted on the right side.

experiments) one of the two populations started outcompeting the other one, which eventually disappeared (for details, see Quer et al. 1996; Solé et al. 1999). The simplified diagram shown in figure 3.4 depicts the evolution process using two coupled landscapes (one for each competitor), each with 3-bit genomes. Initially, both competitors expand in their landscapes, climbing toward the global optimum, but competition eventually selects only one winner (left column).

3.3 Epistasis in RNA Viruses

The two landscapes studied so far from the point of view of evolutionary dynamics define two rather special scenarios. One assumes a rather unrealistic sharp peak with high fitness surrounded by a flat landscape composed by all the rest of sequences displaying equal (lower) fitness. The one presented in the previous section considers a linear decay of fitness as we move away from the peak. But it has been found that, in fact, the shape of the landscape around a fitness peak can sensibly depart from the linear picture.

A key concept to determine the ruggedness of a landscape is *epistasis*, namely the nonlinear interaction between genes in determining a phenotype, in this particular case, the phenotype being fitness. Epistasis is a pervasive concept in evolutionary biology that is central for theories seeking to explain genetic systems such as sex and recombination, dominance, robustness, and the rate of adaptive evolution (de Visser and Krug 2014). Let us first consider a simple case with just two loci, where the wild-type will be AB and the double mutant will be ab. If the combined effect of two mutations in fitness can be fully predicted from the effect of each individual mutation, then no epistasis exists and the expected fitness will be equal to the sum of the fitness effects of the individual mutations (the linear landscape described above would be an example). In terms of

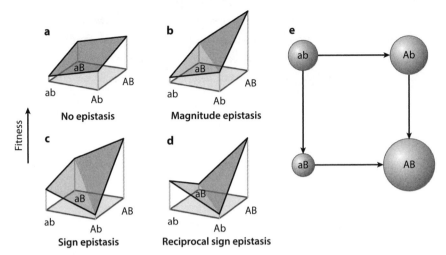

Figure 3.5. Different types of epistases between two loci defining the fitness of a genotype. Capital letters indicate the wild-type and small letters the mutant alleles. (a) In case of no epistasis, the fitness of the double mutant *ab* results from multiplying the fitness effects of both mutations on the wild type genetic background (i.e., the fitnesses of genotypes *Ab* and *aB*). (b) If magnitude epistasis exists, the fitness of the double mutant *ab* is different from the multiplicative expectation. In the example, the observed fitness of *ab* is larger than expected as a consequence of positive epistasis. In the cases both of no epistasis or of magnitude epistasis, the effects of mutations *aB* and *Ab* are unconditionally beneficial. (c) If the effect of one of the mutations is conditionally beneficial (i.e., beneficial in one genetic background but deleterious in another), then we are in the situation of sign epistasis. (d) Finally, if both mutations *aB* and *Ab* are deleterious by themselves, but beneficial when combined, we are in the situation of reciprocal sign epistasis.

fitness landscapes, this situation depicts a smooth surface with no curvature (figure 3.5a). If the observed fitness of a double mutant exceeds the additive effects expectation by a factor of ξ then the two mutations (or genetic loci) had positive epistasis (sometimes called synergistic). Conversely, if the observed fitness of a double mutant is less than the additive effects expectation by a factor of ξ, then the two mutations had negative epistasis (sometimes called antagonistic). This is the so-called magnitude epistases and its effect on the fitness landscape is to introduce a degree of curvature ξ (figure 3.5b; in this particular case, positive).

The situation can be more complicated. Figure 3.5c illustrates the case of sign epistasis, meaning that the sign of a fitness effect depends on the genetic background where it happens. For example, a mutation may have a beneficial given effect one background while being deleterious given another one. Sign epistasis creates ruggedness in the fitness landscapes by introducing ridges and valleys. Ridges represent accessible evolutionary paths, whereas valleys represent low-fitness regions that are not accessible by natural selection. Still, a population can move from the starting to the final point through the ridge. A final situation, depicted in figure 3.5d, is the case of reciprocal sign epistasis, in which the signs of both mutations vary depending on the genetic background where they appear. In the example, both mutations are deleterious by themselves but overcompensate for their effects when pooled together. Reciprocal sign epistasis creates valleys that cannot be traversed by the populations simply by natural selection. Populations get trapped in a peak even if it might not represent the optimal evolutionary solution.

If we return to the peaked landscape view again, one way of depicting the deviations from the linear case is by writing down a generalized fitness function associated to the Hamming distance, defined as the number of single site differences from the wild type. If we assume a given "master" sequence S_m (see chapter 2) and the distance to it, we can write the fitness of a sequence separated mutations from the master as:

$$f(D_h) = 1 - \frac{1}{\nu} D_h^\xi, \qquad (3.18)$$

where the Hamming distance was defined above, and ξ measures the sign and strength of epistasis (Elena et al. 2010). For $\xi = 1$

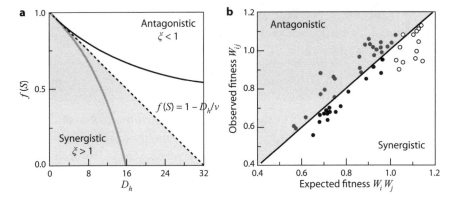

Figure 3.6. (a) Simple fitness landscapes where deleterious mutations lead to a decrease in fitness can be described in terms of the distance to the wild type sequence, following a general functional form $f(S) = 1 - D_h^\xi/v$. Here D_h is the Hamming distance, v is the genome length, and the effect of mutations is introduced by a parameter ξ. Two domains are indicated here, associated to positive or synergistic (gray area, $\xi > 1$) and negative or antagonistic (white area, $\xi < 1$) effects. (b) Evaluation of the pattern of epistasis among pairs of random mutations in VSV. The diagonal represents the case of no epistasis (additive effects).

we have just a linear decay (straight line, figure 3.6a), indicating additive effects. However, when $\xi < 1$, antagonistic epistasis is present, with fitness decaying in a slower way, indicating that the additional mutations have a smaller impact on fitness as mutations are added. This is the so-called antagonistic (positive) epistasis, to be opposed to the so-called synergistic one (for $\xi > 1$), where fitness decays faster as mutations accumulate.

Is it possible to quantify the strength and type of epistasis from experimental data? The answer is positive (Sanjuán et al. 2004; Sanjuán and Elena 2006). Consider the possibility of having both the fitnesses associated to a given set of n loci, namely $W_{1,\dots,i,\dots,n}$. Additionally, consider the fitness of each single mutant, W_i. This number if obtained from experimental data and is defined relative to a wild-type genome. If ρ is the average per cell progeny for the wild-type, and we indicate as r_i and r_0 the respective mutant and wild-type growth rates, W_i is given by (Sanjuán and Elena 2006):

$$W_i = \frac{\rho^{r_i/r_0} - 1}{\rho - 1}. \tag{3.19}$$

If no epistatic effects exist, it should be expected that the overall fitness associated to all loci would be just the product of all W_i terms, consistently with statistical independence. Deviations from such a rule would indicate epistasis. In Sanjuán et al. (2004) and Sanjuán and Elena (2006) the following index is given:

$$\xi_{1,\ldots,i,\ldots,n} = W_{1,\ldots,i,\ldots,n} - \prod_{j=1}^{n} W_j. \tag{3.20}$$

In this way, it was possible to provide a quantitative measure of epistasis. Moreover, this coefficient can be positive or negative, indicating the presence of antagonistic and synergistic epistasis, respectively. When this measure was applied to a number of model organisms, including viruses, bacteria, yeast, molds, and fruitflies, the epistasis coefficient was shown to decay with genome complexity (Sanjuán and Elena 2006). Here simpler organisms (viruses and bacteria) displayed positive values of $\xi_{1,\ldots,i,\ldots,n}$ while higher organisms presented a negative association.

In a systematic analysis of RNA viruses using many genotypes of VSV, each carrying pairs of specific nucleotide mutations, it was possible to use this method to determine the presence of epistatic interactions. The results are shown in figure 3.6b, where observed fitness W_{ij} is displayed against the expected multiplicative value $W_i W_j$. Mutations included both deleterious (filled circles) and beneficial (open circles) effects. The straight line marks the null hypothesis $W_{ij} = W_i W_j$ while deviations indicate the presence of epistatic interactions, i.e., $\xi_{ij} = W_{ij} - W_i W_j \neq 0$. The reported results revealed a very significant role of epistatic interactions, both synergistic and antagonistic. On average, epistasis was pervasive and positive, meaning that the effect of two mutations together was weaker than expected from their independent effects. These results also indicate that, despite the high adaptability of RNA viruses, the presence of antagonistic epistasis between different parts of their genomes can actually impose some barriers on viral adaptation (Holmes 2003).

3.4 Experimental Virus Landscapes

Can we experimentally evaluate the structure and ruggedness of virus fitness landscapes larger than pairs of random mutations? This is a very complex and time-consuming task, but the answer again is positive. Indeed, given the genomic simplicity of viruses, this shall be a much easier task than for other more complex organisms. In recent years, much effort has been devoted to experimentally explore the topography of adaptive fitness landscapes for single molecules and for evolving bacteria adapting to a simple environment (reviewed in de Visser and Krug 2014). In all cases the experimental procedures are similar. Imagine that during an evolution experiment, evolving populations have fixed k mutations. We can precisely identify these mutations by sequencing the genomes of the ancestral and evolved organisms.

Depending on how developed the genetic analyses tools for each particular biological system are, it may be possible to generate all possible 2^k mutant genotypes, that is, all genotypes containing every single mutation individually, all genotypes containing all pairwise combinations of mutations, all genotypes containing all triplets of mutations, and so on until the genotype containing all k mutations. These genotypes describe the full genotypic landscape that corresponds to the adaptive process observed in the laboratory (figure 3.7). In the case of small viruses for which an infectious plasmid containing a copy of the viral genome exists, all these genotypes can be constructed by site-directed mutagenesis, a standard technique implemented in most modern virology laboratories. Finally, the fitness of each

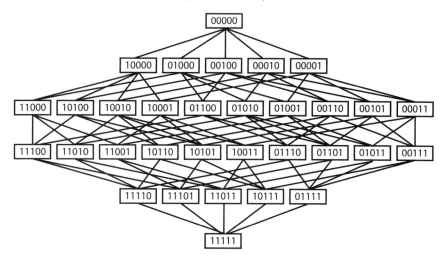

Figure 3.7. Empirical fitness landscape of size $k = 5$. The presence/absence of a given mutation is indicated by 1/0, respectively. Genotypes are ordered top-down starting from the ancestral wild type virus (00000) and finishing with the genotype carrying five mutations (11111). Each row represents genotypes with equal number of mutations. Lines represent potential mutational steps.

genotype in the empirical landscape can be measured to provide a quantitative description of the "height" of the peaks. In the case of RNA viruses, proxies to fitness can be infectivity, virus accumulation, or growth rates that can be expressed relative to the wild-type genotype.

Unfortunately, few empirical studies have addressed the topography of the fitness landscape for viruses. Characterization of the type of epistasis among random pairs of mutations has been done for several RNA viruses (reviewed in Elena et al. 2010). Following the definitions and methods outlined in the previous section, Lalić and Elena (2012) followed a similar approach to evaluate the fitness of a collection of random single and double mutant-generated by site-directed mutagenesis on the genome of *Tobacco etch virus* (TEV), a plant RNA virus. The fitness of all these mutants was evaluated in the natural host tobacco

(*Nicotiana tabacum*). In this case, epistasis was also very common, and of positive sign.

More interestingly, sign and reciprocal sign epistases were pervasive, suggesting that the fitness landscape of this virus was rugged. These two studies, however, explore epistasis among random pairs of mutations, which itself provides very useful information. However, from an adaptive dynamics perspective, it would be much more informative to characterize the ruggedness of the landscape defined by sets of beneficial mutations. To tackle this problem, in a recent set of studies, the landscape defined by the five mutations fixed by TEV during experimental evolution and adaptation to its novel host *A. thaliana* has been thoroughly characterized. All $2^5 = 32$ possible genotypes were constructed and their fitness measured in the novel host (Lalić and Elena 2015). The resulting landscape is shown in figure 3.8a. The topography of this landscape was rugged, defined by prevailing epistatic effects between mutations.

Cases of reciprocal sign epistasis were common among pairs of mutations, defining fitness valleys. The landscape contained two fitness peaks, neither of which corresponded to the genotype containing all five mutations. Interestingly, the landscape contained a whole region of deleterious and lethal genotypes (in red), suggesting that these mutations rose in the population in genotypes that already contained compensatory mutations. Very interestingly, higher-order epistasis was found to be as important as pairwise epistasis in its contribution to fitness. Higher-order epistasis corresponds to interactions between more than pairs of mutations. From a geometric perspective, third-order epistasis represents the effect that a third mutation exerts on the surface defined by a pair of mutations. Likewise, fourth-order epistases quantify the effect that a surface (defined by a pair of mutations) exerts on the surface defined by another pair of mutations.

In a follow-up study (Cervera et al. 2016) the effect of the host species on the ruggedness of the landscape was also analyzed. To do so, the authors evaluated the fitness of the

a

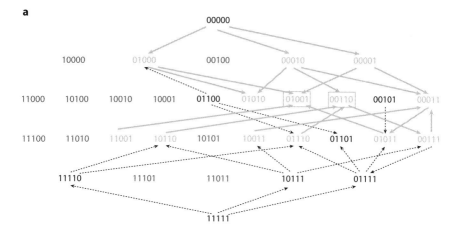

b

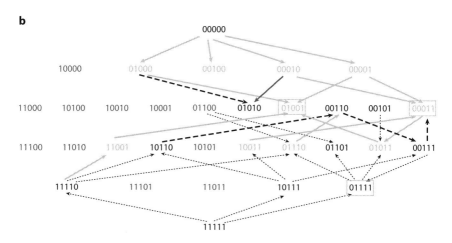

Figure 3.8. Empirical fitness landscapes evaluated for the set of five mutations fixed by TEV during its experimental adaptation to *Arabidopsis thaliana*. The fitness of the 32 genotypes was evaluated in the novel host (a) and in the original one, *N. tabacum* (b). Each string of bits represents a genotype. 1's represent a mutation in the corresponding locus, while 0's correspond to the wild-type allele on this locus. Genotypes in a green box correspond to local fitness peaks. Green lines correspond to beneficial mutations, genotypes in red are deleterious, and black dashed lines correspond to neutral changes (in the direction from wild-type to the quintuple mutant genotype).

same collection of TEV mutants, but with the original host, *N. tabacum* (figure 3.8b). Two important conclusions can be drawn from this study. First, the topography of the landscape was smoother in the novel host than in the ancestral one. This has important evolutionary implications, for example, exploration of the landscape is more efficient in the novel host. Second, both landscapes were uncorrelated, meaning that genotypes that were beneficial in the novel host were deleterious in the ancestral one. This is the typical situation of antagonistic pleiotropy that will be discussed in chapter 6 in the context of emerging viral infections. Differences between the two landscapes, however, were local rather than global, with particular genotypes changing their relative height in the landscape and resulting in different patterns of epistatic interactions with their neighbors.

This dependence of the topography of the fitness landscape on the host supports the notion of dynamic landscapes rather than of static ones. Nonetheless, both landscapes shared common features, such as the existence of fitness holes due to unconditionally lethal genotypes or the presence of pervasive epistatic interactions. The topography of both empirical landscapes matches pretty well with the expectations from a random uncorrelated landscape; lying somewhere between the extreme case of the House-of-Cards model (Kingman 1987), in which the fitness of each genotype is absolutely independent of the fitness of the other genotypes, and the less radical case of the rough Mount Fujimori model (Aita et al. 2000), which combines properties of both the House-of-Cards and a purely multiplicative landscape.

3.5 The Survival of the Flattest Effect

The existence of a quasispecies strongly affects the way selection acts, because the evolutionary success of individual genomes depends not only on their own replication rate but also on the average growth rate of the quasispecies they belong to.

Mutation acts as a selective agent that shapes the genome in a manner that causes the entire quasispecies to become robust against mutations, and fast replicating genomes that produce low-fitness offspring can be outcompeted by slow replicating genomes provided the latter quasispecies inhabits a region of sequence space characterized by high neutrality and connectivity (Schuster 1988; van Nimwegen 1999; Wilke 2001b). This phenomenon has been dubbed the "quasispecies effect" (van Nimwegen 1999; Wilke 2001a) or more recently as "the survival of the flattest" (Wilke et al. 2001), in clear reference to Darwin's "survival of the fittest" concept. Indeed, authors who cast doubts about the relevance of the model to real viruses based their criticism on the fact that the quasispecies effect was not observed in vivo (Holmes 2002).

However, a recent experiment gave support to the validity of the "quasispecies effect" in real viral populations. In one of them, Codoñer et al. (2006) selected two different viroids that infected a common host. These two viroids largely differ in their replication rates and in the extent of genetic variability they generate within their host. CChMoVd generated lots of variants after being inoculated but accumulated to very low titers. *Chrysanthemum stunt viroid* (CSVd) accumulated to very high titers but showed little genetic variation. The authors hypothesized that CChMoVd may represent the case of a flat organism replicating in a neutral network (NN) whereas CSVd may not. To test this hypothesis both viroids were co-inoculated into the same plants and allowed to compete. As expected, CSVd quickly outcompeted CChMoVd owing to its faster replication rate. However, when the mutation rate was artificially increased by UVC radiation, the situation was reversed and CChMoVd persisted in the mixed population.

In another experiment, Sanjuán et al. (2008) provided additional evidence of the survival of the flattest effect. Two VSV populations that differed in evolutionary history were chosen. Population *A* was formed by individuals that on average had lower fitness than those from population B but that were more diverse in fitness. The authors hypothesized that population A was the flattest while population B was the fittest. As in the viroids case, these two populations were allowed to compete in standard conditions and at increasing mutation rates (in this case by adding either one of two chemical mutagens, 5-FU or 5-AzC). The results showed that while population B outcompeted population A under standard conditions, B was able to reverse its fortune as the concentration of mutagens increased.

A minimal mathematical model can be used to describe the survival of the flattest effect (figure 3.9e-f). We use a mean field model to analyze the dynamics between two quasispecies located in two different (but coupled through competition) fitness landscapes. The state of the system is thus described by a state $\Gamma = (x_0, x_1, y_0, y_1)$ such that $\sum_i (x_i + y_i) = 1$ (constant population constraint). The model is given by the next four-dimensional dynamical system, with two equations for the "fit" populations reading

$$\frac{dx_0}{dt} = r_0 Q x_0 - x_0 \Phi(\Gamma) \tag{3.21}$$

$$\frac{dx_1}{dt} = r_0(1 - Q)x_0 + r_1 x_1 - x_1 \Phi(\Gamma), \tag{3.22}$$

which include a Swetina-Schuster approach (see chapter 2) with $r_0 > r_1$ and associated with a "sharp" peak. These two equations are coupled with the other two by means of the competition terms including the $\Phi(\Gamma)$. The "flat" component now involves two additional equations,

$$\frac{dy_0}{dt} = r_0 Q y_0 + r_1(1 - Q)y_1 - y_0 \Phi(\Gamma) \tag{3.23}$$

$$\frac{dy_1}{dt} = r_1 Q y_1 + r_0(1 - Q)y_0 - y_1 \Phi(\Gamma), \tag{3.24}$$

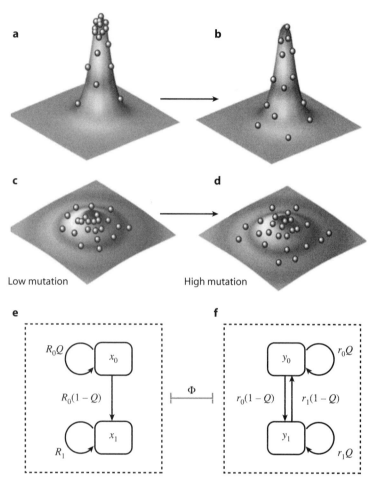

Figure 3.9. The fittest versus the flattest. Here two idealized landscapes are indicated with their populations (gray spheres) associated to the peaks. For the sharp peak (a-b) at low mutations, genomes are close to the peak whereas increasing mutation pushes them far toward low fitness values. For a flat landscape (c-d) the situation is different, since the spread has little impact on average fitness. A minimal model of this phenomenon can be described in terms of a simple quasispecies model (e-f).

that allow back mutations and thus a diffusion-like coupling. To make things simpler, we will asume that $r_0 = r_1 = r$, and additionally we will have $r_0 > r > r_1$, meaning that the flat strain has an intermediate fitness value between the maximum and the minimum of the fit one.

Given the homogeneous replication rate r, the two "flat" components will be undistinguishable, and it is easy to see that the two last equations can be collapsed into a single one using $y = y_0 + y_1$, namely

$$\frac{dy}{dt} = ry - y\Phi(\Gamma). \qquad (3.25)$$

The analysis of the resulting three-species system (now we have $\Gamma = (x_0, x_1, y)$) gives a condition for the survival of the flattest:

$$Q < \frac{r}{r_0}; \qquad (3.26)$$

or, using the mutation rate $\mu = 1 - Q$, if the critical condition

$$\mu > \mu_c = Q < 1 - \frac{r}{f_0} \qquad (3.27)$$

is met. We thus have a well-defined condition connecting the relative distance between the fitness of the highest peak and the one associated to the flat strain. The dynamics under the two different regimes can be better appreciated by using a spatial model (Sardanyés et al. 2008). In figure 3.10 we illustrate the outcome of competition under low (a) and high (b) mutation rates in a cellular automaton model where four states are used. The lighter areas in both pictures correspond to the fittest sites, which propagate at the expense of the flat states until the lattice is filled. Conversely, the darker sites associated with the flat states are the winners in the second case. Note that the expansion process takes longer in the second scenario, in accordance with the lower replication rate of the less-fit species.

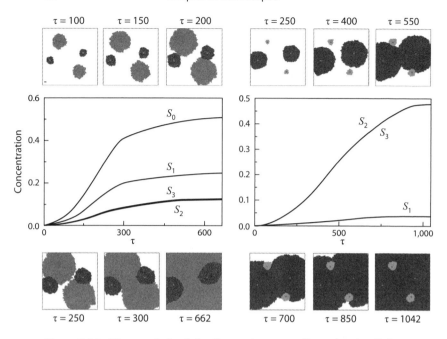

Figure 3.10. The survival of the flattest in a two-dimensional cellular automaton model (Sardanyés et al. 2008). The left and right diagrams and snapshots are obtained for low and high mutation rates.

3.6 Virus Robustness

In the previous section, we discussed a particularly interesting point, namely, selection may favor not the fittest (i.e., faster replicators) but the flattest (i.e., more robust replicators), which better tolerate the accumulation of deleterious mutations. Theoretically, viruses with high mutation rates may be favored in stressful environmental situations where the input of beneficial mutations allows for escape and survival (e.g., changing cell types, tissues, and hosts, or the presence of antiviral responses or drugs). However, in all situations, deleterious and lethal mutations represent the larger fraction of all possible mutations, thus jeopardizing viral fitness. How do RNA viruses maintain their functionality in this scenario? Are they robust against the accumulation of deleterious mutations? Is the survival of the flattest the only mechanism they may enjoy to buffer the effect of mutations? In a hallmark article, de Visser et al. (2005) reviewed the notion of robustness and explored its causes and consequences. Robustness is the preservation of the phenotype in the face of perturbations. The robustness of phenotypes appears at various levels of organization: from gene expression, protein folding, metabolic flux, physiological homeostasis, and development, to fitness. Three reasons may explain the evolution of mutational robustness. Firstly, as long as it is heritable, shows variability among individuals, and affects fitness, mutational robustness can be a selectable trait. The more frequent mutations are, the stronger will selection be at promoting the evolution of mutational robustness. Secondly, mutational robustness is a side effect of stabilizing selection acting on different traits. Thirdly, given that environmental fluctuations often have a strong impact on fitness, selection would favor mechanisms of environmental robustness, and then mutational robustness simply evolves as a side effect (dubbed plastogenetic congruence). This last explanation is particularly appealing for RNA viruses as they must cope not only with deleterious mutations but also with dramatic and fast fluctuations in their environments.

Elena et al. (2004) elaborated on the possible mechanisms by which RNA viruses may attain mutational robustness, distinguishing two classes of mechanisms. Mechanisms of intrinsic robustness are the consequence of RNA-genome architecture, replication peculiarities, and population dynamics. Intrinsic robustness mechanisms operate efficiently at the population level. By contrast, extrinsic robustness results from the exploitation of cellular buffering mechanisms by viruses.

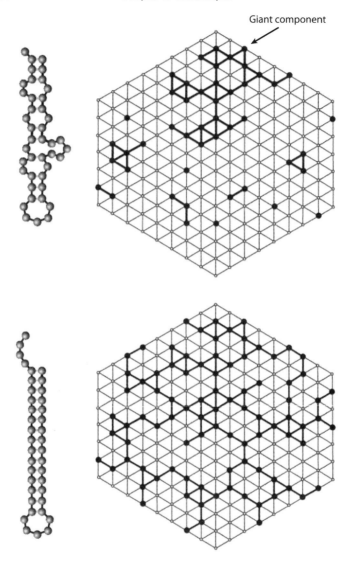

Figure 3.11. Neutral networks in sequence space. Two idealized examples of the neutral networks connecting sequences that lead to the same
type of fold (adapted from Schuster (2010)). As a result of neutrality, the
landscape of RNA folds associated with different RNA sequences displays
islands of connected neutral phenotypes that can percolate through
sequence space.

3.6.1 Intrinsic Mechanisms of Mutational Robustness

In the previous sense, the survival of the flattest discussed above can be considered as an *intrinsic* mechanism of robustness. It depends on selection, pushing viral populations toward regions of genotypic space in which mutations may have a more neutral effect. These regions of high neutrality, or NNs (figure 3.10), allow the virus to accumulate mutations without suffering fitness losses. A positive consequence of the existence of such NNs is that viral populations may efficiently explore distant regions of sequence space.

The existence of such NNs has a strong implication on the antigenic evolution of *Influenza A virus* (IAV) H3N2. The observed patterns of epochal antigenic evolution of H3N2, alternating periods of phenotypic stasis punctuated by sudden changes in the antigenic phenotype, can easily be explained in terms of NNs. At the onset of an epochal evolution cycle, an H3N2 population is distributed over the NN of an antigenic cluster. Neutral mutations accumulate, allowing the virus to explore and reach distant regions of this particular NN. At some point, a mutation at the edge of the network will create an individual that belongs to a new NN that corresponds to a different antigenic cluster. This antigenic innovation corresponds to an epidemiological peak in the number of infections. The new antigenic variant now starts exploring the new NN, and the process repeats itself (Koelle et al. 2006; Van Nimwegen 2006).

A second mechanism of mutational robustness is high ploidy. Viruses are *n*-ploid organisms; *n* is variable during infection. At initial stages, multiplicity of infection (MOI) is low and viruses are effectively haploid. However, as infections progress, high MOIs ensure frequent co-infections and increasing ploidy. An immediate consequence of polyploidy is genetic complementation. Strong complementation slightly reduces the average population fitness by weakening the efficiency of purifying selection but significantly enhances population diversity and mutational robustness, especially if epistasis among deleterious mutations is antagonistic.

Different modes of genome replication may also affect mutational robustness. The stamping-machine replication strategy uses always the same molecule as template for producing all the progeny, thus minimizing the number of mutations at the cost of being slower than a geometric replication strategy. Geometric replication is faster, but has the drawback of using progeny genomes as templates, and thus generating offspring with a number of mutations that increases geometrically. Furthermore, it has been shown that, in combination with selection, the stamping machine accumulates less mild-effect mutations than geometric replication. Indeed, the difference between both replication schemes in terms of minimizing deleterious mutational load is enhanced if mutations show negative epistasis.

A final mechanism of intrinsic mutational robustness is viral sex, resulting from recombination of homologous molecules or in segregation of segments in a multipartite genome. Sex recreates mutation-free genotypes and helps to keep the average population fitness high. Both forms of sex are common among RNA viruses.

3.6.2 Extrinsic Mechanisms of Mutational Robustness

Prokaryotes, archaea, and eukaryotic cells all have a set of proteins that respond to stress, the best known ones being those responding to thermal stress. These heat-shock proteins (HSP) operate as molecular chaperones assisting other proteins to fold properly into their active structures as they are synthesized under thermal stress (Borges and Ramos 2005). In addition to their activity under exogenous stress conditions, it has also been shown that HSP chaperones assist in the folding of mutated proteins, buffering the effect of mutations on the structure, even in the absence of thermal stresses (Queitsch et al. 2002). Indeed, HSPs are considered as capacitators of evolution because they maintain genetic variability hidden from the action of purifying selection; this variation may become visible under stress (e.g., an environmental change) when the chaperones are diverted into their original function (Queitsch et al. 2002). Therefore, chaperones can be seen as robustness machines.

A well-established observation is that viral infections induce the cellular stress response, including the over-expression of many members of the HSP family (Jacob et al. 2017; Neckers and Tatu 2008). Very interestingly for the discussion in hand, it has been shown that most viruses not only induce HSPs' expression as a result of cellular stress, but also need cellular chaperones during their life cycle to solve their own protein-folding problems (Aparicio et al. 2005; Geller et al. 2012; Low and Fassati 2014), to assist during RNA replication, and to interfere with cellular processes such as signal transduction. Therefore, it is tempting to suggest that viruses are hijacking the cellular HSPs for their own benefit as a way of buffering mutations that otherwise may have a negative impact on their fitness due to destabilizing effects on protein structures.

3.7 Selection: Fitness versus Robustness

A mathematical model provides a consistent picture of this selection for robustness associated to the presence of neutrality (Wilke 2001a). Although this model is somewhat related to the mean field model discussed above, it explicitly incorporates neutrality as a system's parameter and allows us to make predictions that can be explicitly defined and related to relevant variables such as genome size.

Consider first one neutral network $\Omega \subset \mathcal{H}^\nu$ and the probability Q that a mutant string S' is generated from $S \in \Omega$ (Ofria and Adami 2001). To do this calculation, a parameter η is introduced, giving the probability that a single site mutation has

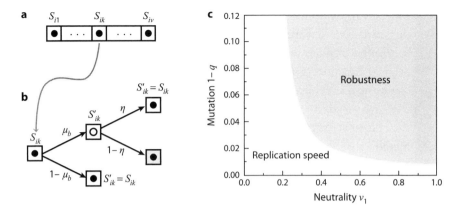

Figure 3.12. Modeling the interplay between replication and neutrality. Given a digital genome (a) the probability Q of a neutral mutation is estimated from the two favorable events indicated in (b). In (c) the two phases consistent with either selection for spread or selection from robustness are shown.

no phenotypic effect. If μ_b is the mutation rate per site, we have (see figure 3.12a):

$$Q = [1 - \mu_b(1 - \eta)]^\nu \approx e^{-\mu_b(1-\eta)\nu}. \qquad (3.28)$$

This calculation can now be used to develop a first population dynamics model where x_1 and x_d will stand for the concentrations of genomes respectively within the neutral network Ω and outside of it. The model thus uses $x_1 + x_d = 1$ and the dynamics is given by:

$$\frac{dx_1}{dt} = r_1 Q x_1 - x_1 \Phi(x_0, x_d) \qquad (3.29)$$

$$\frac{dx_d}{dt} = r_1(1 - Q)x_1 - x_d \Phi((x_0, x_d). \qquad (3.30)$$

As defined, the model assumes that the sequences outside Ω are unable to replicate and using this assumption the equilibrium concentration is

$$x_1 = Q = e^{-\mu_b(1-\eta)\nu}. \tag{3.31}$$

The importance of this result is that knowing how x_1 changes as μ_b grows it is possible to estimate the network's neutrality, which follows this prediction accurately even when sophisticated, molecular-level frameworks are used (Wilke 2001a).

Finally, the conditions for selection for replication speed versus those for robustness will be obtained by analysing a three-population model where two populations are associated to two different neutral networks and again the one for genomes outside both networks is included:

$$\frac{dx_1}{dt} = r_1 Q x_1 - x_1 \Phi(x_1, x_2, x_d) \tag{3.32}$$

$$\frac{dx_2}{dt} = r_2 Q x_2 - x_2 \Phi(x_1, x_2, x_d) \tag{3.33}$$

$$\frac{dx_d}{dt} = r_1(1 - Q)x_1 + r_2(1 - Q)x_2 - x_d \Phi(x_1, x_2, x_d), \tag{3.34}$$

where the condition $x_1 + x_2 + x_d = 1$ leads to $\Phi = r_1 x_1 + r_2 x_2$. The analysis of this system leads to a condition for the critical mutation rate separating the two selection phases:

$$\mu_c = 1 - \frac{\ln(r_2/r_1)}{\nu(\eta_1 - \eta_2)}. \tag{3.35}$$

Two possible steady states exist (obtained from $dx_k/dt = 0$, $k = 1, 2, d$). These are $(Q_1, 0, 1 - Q_1)$ and $(0, Q_2, 1 - Q_1)$. We can see them drawn in figure 3.12b. As the figure illustrates, in one, the successful strategy for the population x_1 is to be a fast replicator, whereas for mutation rates larger than the critical value, the successful strategy requires neutrality.

The landscape picture helps to conceptualize the idea of evolution of a population of genomes and to include relevant properties (such as neutrality of ruggedness) in an explicit manner. The previous definitions and approximations provide a general background to illustrate the key concepts associated to the genotype-phenotype mapping. Formal generalizations have been proposed (Stadler 1999; Stadler et al. 2001; Reydis and Stadler 2002), as well as extensions that incorporate a multiscape picture (Catalán et al. 2017). But perhaps the most crucial missing element here is that viruses have evolved in a host context where both partners are forced to change. To understand the nature of viruses, we need to put both hosts and pathogens at play.

4

VIRUS DYNAMICS AND ARMS RACES

4.1 Virus-Host Interactions

The life histories of viruses cannot be understood outside the context of their host partners. Because they cannot exist without the environment defined by their target organisms, understanding the behavior of viruses inevitably leads to consideration of how they and their hosts coevolve. In the previous chapters we have already discussed the structural, computational and ecological features associated to viruses. But the whole picture would be incomplete unless we take into account the coevolution of viruses and their hosts. As stated by Theodosius Dobzhansky, "nothing makes sense unless under the light of evolution," and this is particularly true for viruses, which are rapidly evolving entities with an enormous impact on the evolutionary history of their hosts; and, as we will see, the converse holds too. Moreover, the lessons learned from their study have also shed light on several important approaches to both applied biomedical and basic research.

Mathematics is also key to understanding evolution. Charles Darwin himself wrote once that he "deeply regretted that I did not proceed far enough at least to understand something of the great leading principles of mathematics." Darwin perceived that a formal picture of natural processes gave their users "an extra sense" (May 2004). At the time Darwin wrote this, physics was already a well-established and powerful discipline, and its power would only increase in the next two centuries thanks to the development of mathematical models capable of astonishingly accurate predictions. Biology is seen by many as a much more messy and qualitative field, but the development of population genetics and theoretical biology and the rise of systems biology at the end of the twentieth century rapidly changed these views. In this chapter we will study several models describing key aspects of virus-host dynamics on different scales. As an illustrative example, we will focus our discussion on the interaction between HIV-1 and the immune system, using both ecological and evolutionary models. Such modeling is, in principle, a highly challenging goal. The immune system is far from simple, and a serious consideration of the relevant biology rapidly ends up in a clumsy, multi-parametric model that is of little use in terms of insight. This is one of the problems faced by complex systems practitioners: what to ignore as nonessential and what to include in the model description.

Because multiple players are involved, from specific molecular complexes to a high diversity of cell types, modeling these interactions is a difficult task (Perelson and Weisbuch 1997; Molina-Paris and Blythe 2011). One key step in defining a set of equations or rules is deciding which details should be included and which should be left aside. This choice implies some degree of arbitrariness, and the guiding principle must be. "What is my question?" This is not a minor point. Oftentimes, models can look too abstract to specialists (usually trained in a highly specialized area) since they ignore biologically relevant features of the "real" biology. However, it is not less true that real insight into many difficult problems requires some degree of abstraction and some simplicity (Gell-Mann 1994; Solé and Goodwin 2001; Mitchell 2012).

The problem is illustrated by the diagram depicted in figure 4.1a, where we show a small part of the interactions between immune system defense mechanisms related to HIV-1 infection dynamics (Walker and Yu 2013). Both the innate and adaptive immune responses are included, involving several classes of specialized cells as well as complex molecular signals. This defines a network of interactions involving several scales in terms of the nature of biological players, but also different temporal and spatial scales related to fast responses as well as long-term memory dynamics. In this chapter we will not describe the organization of immune responses, which include different classes of response mechanisms (see, for example, Janeway 2001) and would require a lengthy presentation. Only some specific features and players will be taken into account and introduced as needed. Let us mention only that any foreign agent (such as a virus) can be detected, ingested by some cells, and broken into pieces. Each one of these foreign fragments acts as a specific antigen that can be recognized by another class of cells in specific ways. When this happens, the cell population grows and expands rapidly, and results in the removal of all cells marked with that specific marker. If the virus does not mutate fast, this immune response can get rid of the virus. If the virus is able to generate "escape" mutants, the process starts again.

Similarly, the life cycle of HIV-1, which involves several key steps associated to attachment to cell surface, reverse transcription, integration in the host's genome, and maturation of newly produced viral particles (figure 4.1b), will not be explicitly considered. However, the knowledge of these molecular processes allows us to introduce specific modeling rules that explain the dynamics of viral propagation and how it is affected by drugs targeting some of these steps. Here mathematical models can help us gain an understanding of the tempo and mode of viral evolution and its impact on disease progression. The HIV-1 case is a wake-up call on the views of viruses as simple parasites

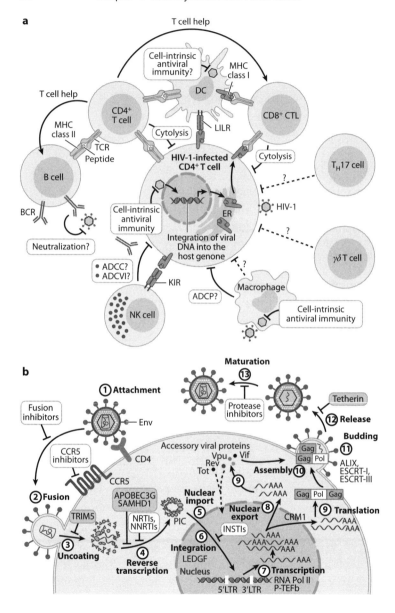

Figure 4.1. Continued caption on next page.

that replicate and spread. When it had started to be known by the early 1980s, the clinical syndrome associated to the profound depression of immune responses was named AIDS, for *acquired immunodefficiency syndrome*. Among other things, patients exhibited a marked immune system impairment favoring secondary infections including pneumonia and a rare form of cancer known as Kaposi's sarcoma. Eventually, the virus was isolated and its small genome sequenced and studied in full detail (see chapter 1).

But understanding its life cycle, how it infects its target cells, where it came from, or how it evaded immune responses and drug therapies took a long time. Such understanding required dedicated efforts by thousands of scientists and clinicians and was slowly achieved, while millions of human beings became victims of the disease. Along with these efforts, mathematical models helped in solving some of the fundamental questions concerning the work of this deadly molecular machine.

4.2 HIV Multiscale Dynamics

It was soon found that HIV-1 was a retrovirus, using (and encoding for) a reverse transcriptase enzyme, capable of synthesizing DNA using as a template the two genomic RNAs encapsulated within each viral particle. This DNA can then be read by the cellular machinery, creating multiple copies of the virus' genomic RNA, but also (and crucially) it can get integrated into the

Figure 4.1. Immune response complexity and HIV-1 infection dynamics. Two levels of complexity are summarized here. The diagram (a) depicts many (but not all) relevant interactions between the HIV-1 and our immune system (Walker and Yu 2013). Most models just consider specific pairwise interactions. Similarly, many crucial features of HIV-1 have to do with its life cycle inside target cells (b) with several key molecular events involved. Each step has been targeted by specific antiretroviral drugs. Adapted from Laskey and Siciliano (2014).

host genome (Bushman et al. 1990; Farnet and Haseltine 1990) by means of another viral protein named integrase. Integration happens almost randomly in the host genome, with millions of unique integration sites being described (Serrao and Engelman 2016). This integrated virus can remain in a latent stage for long periods of time, protected from the cell's innate immunity defenses, or can constantly be transcribed to produce new genomic RNAs and, hence, new viral particles that are excreted outside the cell. An advantage of this hiding strategy is that the integrated virus now can spread vertically when infected cells multiply. Additionally, the target of the virus has also a special status: HIV infects a class of immune cells known as CD4$^+$ T cells, which are a particular class of lymphocytes also known as T helper cells. The virus binds to the CD4 receptors, expressed on the surface of these cells. Since the virus infects and can remain latent within cells that are actually a key part of organismal defenses, we can appreciate the potential harm that can result from it.

The interaction between this retrovirus and the human immune system provides an excellent (and unfortunate) example of the multiple scales involved in the evolution, propagation, and treatment of HIV-1. In the next sections, we will consider several examples of models spanning several scales, all of them relevant for different reasons and all presenting important challenges. Beyond the special idiosyncrasies associated with each scale, one central observation pervades most of what we nowadays know about HIV-1. This is the anomalous clinical observation concerning the collapse of the immune system of the patient summarized in figure 4.2. Here the viremia (measured as number of copies of HIV-1 genomic RNA molecules in a blood sample) and the estimated concentration of CD4$^+$ T cells are plotted over a time span of more than one decade. The diagram starts from the early infection (showing a marked rise of viremia and flu-like symptoms), all the way to the eventual appearance of AIDS

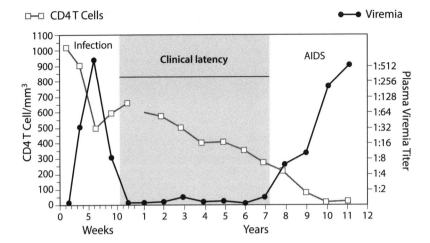

Figure 4.2. The three-phase dynamics displayed by most patients infected with HIV-1 (adapted from Pantaleo et al. (1993)). The two key variables, namely HIV-1 concentration (in RNA copies per volume of blood) and the concentration of its target cells (CD4$^+$ T cells), are displayed. The phases (from left to right) are a brief acute phase, the asymptotic period (highlighted in gray), and the AIDS phase.

symptoms. These symptoms are tied to a high viremia along with a marked decay in immune cell counts (Pantaleo et al. 1993). In order to see clearly the progression, the time scale is shifted from weeks to years at the center of the plot. What we can clearly appreciate is that a puzzling decay of viral counts is present right after the original peak and that viral populations remain very low over several years, with little or no external clinical signs of a real problem taking place.

The latency revealed by these data creates an apparent paradox. If HIV-1 is responsible for the development of AIDS and is truly efficient at killing its target CD4$^+$ T cells, one should expect to see a growth in the number of viral particles along with a constant supression of CD4$^+$ T cells until collapse arises. A reduced number of viruses instead suggests that they are kept under control by the immune system and that the emergence of AIDS might have to do with other factors not directly related to HIV-1. Some authors even claimed that AIDS had nothing to do with HIV-1 and could even be a by-product of antiretroviral drugs.[1] To solve the puzzle, a combination of both experimental and mathematical approaches is required.

4.3 Population Dynamics of HIV Infection

Take the following example. When a model of predator-prey interactions is to be built, population biologists usually consider the number of individuals in each of two species (say X and Y for preys and predators, respectively) as the key variables. Interaction, growth, and mortality parameters are then introduced and the main problem is how to properly express the functional relations associated to all these processes (Case 2000). These types of models often map interactions into some class of "reactions" among components (figure 4.3 a-b): a predator finding a prey can be reduced to a particle of a given class Y "colliding" with a particle of class X. As a result, with some probability, the X particle gets removed while a new copy of the Y particle emerges in its place. Moreover, both types of particles "die" with given probabilities, leaving some empty space available for other particles. The list of reactions required to represent this so-called Lotka-Volterra model (May 1973; Gotelli 1988) is given in figure 4.3b. One could say that this is too simple to represent the true complexity of a natural predator-prey system, but the truth is that this toy model accounts for one particularly relevant property of these systems, namely the presence of cycles, which are a consequence of the internal dynamics of the system. Instead

---

[1] Such an unfortunate kind of claim, along with irresponsible political decisions grounded in wrong assumptions, extreme religious positions, and ignorance have to be blamed for much of the spread of the pandemics.

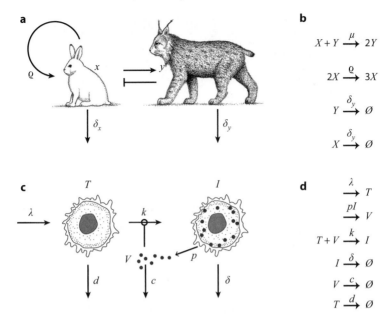

a

b

$$X + Y \xrightarrow{\mu} 2Y$$

$$2X \xrightarrow{\varrho} 3X$$

$$Y \xrightarrow{\delta_y} \varnothing$$

$$X \xrightarrow{\delta_y} \varnothing$$

c

d

$$\xrightarrow{\lambda} T$$

$$\xrightarrow{pI} V$$

$$T + V \xrightarrow{k} I$$

$$I \xrightarrow{\delta} \varnothing$$

$$V \xrightarrow{c} \varnothing$$

$$T \xrightarrow{d} \varnothing$$

Figure 4.3. Simple models and complex interactions. The population dynamics of (a) predator-prey interactions can be represented as a set of reactions taking place among two types of particles X and Y, as indicated in (b). These simple interactions generate oscillations, consistently with those observed in real ecosystems. Some similarities exist between the interactions occurring between viruses and CD4$^+$ T cells, but some important differences exist too, as shown by the diagrams (c, d) (drawings by R. Solé).

of being driven by some external driver, the nonlinearities of the model fully account for the emergence of oscillatory behavior.[2]

Consider now the problem of a minimal model of viral infection (figure 4.3 c-d), where at least three variables are needed: the

[2]This is not a realistic model, in particular because it generates oscillations with an amplitude that is dependent upon the initial condition. It is in fact a *Hamiltonian* system, where some conservation laws can be defined that are meaningless for real systems. However, the basic prediction is reproduced by a wide class of more realistic Lotka-Volterra models (Case 2000).

number of target cells (T), the number of infected cells (I), and the number of viruses (V). This model has been studied in detail (Perelson and Nelson 1999; Statford et al. 2000; Perelson 2002; Nowak et al. 1996) and assumes that a population of cells is generated at some constant rate λ. These cells are the target of a virus that infects them at some rate k, and this process generates infected cells, which produce virus particles at some rate p. All the three populations decay with given rates (indicated as d, δ, and c in the picture).

The set of interactions also involves (as in the Lotka-Volterra model) pairwise "reactions," but in this case viruses transform their host (turning T cells into infected I cells) into a different type of "particle," and moreover they are produced by the infected population at a constant rate. Additionally, the model assumes that we have a well-mixed system where encounters between viruses and cells occur at random with homogeneous rates. This is of course another simplification: although we have been referring to the clinical data in terms of concentrations in blood samples, the organization of the immune system is far from a liquid one. More importantly, the model considers no evolution of the viral component, which of course is a very strong assumption.

The following set of equations describes the basic dynamics outlined in figure 4.3c-d:

$$\frac{dT}{dt} = \lambda - dT - kVT \qquad (4.1)$$

$$\frac{dI}{dt} = kVT - \delta I \qquad (4.2)$$

$$\frac{dV}{dt} = pI - cV. \qquad (4.3)$$

Let us first consider the simpler scenario where no infection has taken place, and thus $I = V = 0$. A single differential equation is now all we need, namely

$$\frac{dT}{dt} = \lambda - dT; \qquad (4.4)$$

which has a predicted steady state (obtained from $dT/dt = 0$) given by[3]

$$T_s = \frac{\lambda}{d}. \qquad (4.5)$$

Such a steady state prediction seems consistent (at least qualitatively) with the observed viremia of untreated patients (Ho et al. 1995), as shown in figure 4.4a, where the measured concentrations on three patients displaying AIDS are depicted. Despite some fluctuations, we can see that the numbers remain rather constant over time.

The nature of the asymptomatic phase became a puzzle for AIDS researchers and represented a major problem in the understanding of the disease progression. What was the real nature of the process? Since the latency period is characterized by very low viremia, one potential possibility was that the viral population reached some kind of steady state close to extinction. A whole generation of new models of HIV-1 dynamics was developed in the early 1990s that allowed, in close contact with a rising number of clinical data, to uncovering—among other things—the nature of the asymptomatic phase.

The full model gives further intuition concerning the potential outcomes of the infection dynamics. This model has two alternative steady states, obtained from $dT/dt = dI/dt = dV/dt = 0$. The first corresponds to the virus-free system, and is given by

$$S_0 = \left(\frac{\lambda}{d}, 0, 0\right), \qquad (4.6)$$

---

[3]The equation can be also solved by integrating it after separation of variables, i.e., by solving $\int_{T_0}^{T} dT/(\lambda - dT) = t$, which gives a solution $T(t) = T_s - (T_s - T_0 \exp(-dt))$. The steady state T_s is also achieved when $t \to \infty$.

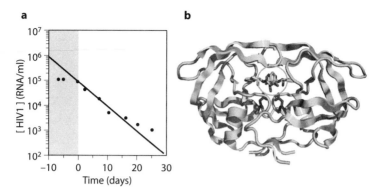

Figure 4.4. Time series of dynamical decay of HIV-1 in plasma (a) as obtained from samples from one patient (among 20) having protease inhibitor treatment (Ho et al. 1995). Treatment starts at time zero ($t = 0$). Prior to treatment (gray area) RNA levels of the virus display steady levels over time. The plot is shown in linear-log scale and the straight line is consistent with an exponential decay of the virus over time. In (b) the 3D structure of the protease is shown, along with the location of the inhibitor (gray circle).

whereas the second is the steady state $S_1 = (T_s, I_s, V_s)$, where we have a steady infected population

$$I_s = \frac{1}{\delta}\left(1 - \frac{dc\delta}{pk}\right) \tag{4.7}$$

and the equilibrium value for the virus population given by

$$V_s = \frac{p}{\delta c} - \frac{d}{k}. \tag{4.8}$$

The last expression gives a critical condition for the virus to persist: since we need $V_s > 0$ to meet this condition, we have $p/\delta c > d/k$ or, by rearranging terms,

$$R_0 = \frac{pk}{\delta cd} > 1. \tag{4.9}$$

This parameter gives a threshold condition for infection success (see chapter 5).

Given the multiple parameters associated to the model (despite the great number of simplifications made), it would be unrealistic to expect all these numbers to be estimated from real data. But some crucial parameter, such as the clearance rate c of the virus, can be easily estimated. And in doing so, it is possible to compute the rate of virus production before therapy begins. A minimal mathematical model can be formulated (Perelson and Nelson 1999) and applied to clinical data from control patients carrying the HIV-1 virus. A therapy was applied involving ritonavir, a protease inhibitor that repressed the synthesis of the virus (Ho et al. 1995; Wei et al. 1995; Perelson et al. 1996) (figure 4.4b). The outcome of this drug trial is shown (for one of the patients) in figure 4.4a. Here the viral load of HIV-1 (as usual, measured in number of RNA molecules/ml) is represented, before (dashed area) and after the treatment with ritonavir (Ho et al. 1995). The observed time series can be explained using a basic dynamical equation for HIV-1, where a production-decay pair is at work, i.e.,

$$\frac{dV}{dt} = P - cV, \qquad (4.10)$$

where it is assumed that the virus is produced at some constant rate P and disappears from the system at a rate c. The constant production term is reasonable under the previous set of assumptions (and always operating under a given time scale). If a protease inhibitor is used, it is possible to observe an exponential decay. On a linear-log plot, this corresponds to a linear function $\log V(t) = \log V(0) - ct$ of HIV-1 load.

The exponential decay is easily explained by using $P = 0$ (i.e., no production due to treatment). The previous equation is now simply a decay model,

$$\frac{dV}{dt} = -cV, \qquad (4.11)$$

with a standard solution given by

$$V(t) = V_0 e^{-ct}, \qquad (4.12)$$

V_0 indicating the initial viral load. A key quantity of this dynamics is given by the so-called half-time parameter τ, which is defined as the time required to halve the initial population. Using this definition, τ is obtained from the condition

$$V(\tau) = \frac{V_0}{2} = V_0 e^{-c\tau}, \qquad (4.13)$$

which leads to measurable time (in days), namely:

$$\tau = \frac{\ln 2}{c} = 2.1 \pm 0.4. \qquad (4.14)$$

In other words, from this simple experiment we can have an estimate of the rate of virus clearance under untreated conditions. Now, let us go back to the steady behavior of virus concentration before antiretroviral therapy (Ho et al. 1995; Perelson and Nelson 1999). We have again the assumption

$$\left(\frac{dV}{dt} \right)_{t=0} = P - c V_0 \approx 0, \qquad (4.15)$$

which gives a relation between production, clearance, and viral load:

$$P \approx c V_0. \qquad (4.16)$$

Since both c and V_0 are estimated, this leads to a virus production rate

$$P \approx 0.7 \times 10^9 \qquad (4.17)$$

virions per day.[4]

This was a crucial result for several reasons and a great success story for modeling (Perelson 2001). First, the results indicated

---

[4]This is a lower bound of the HIV-1 production (Perelson and Nelson 1999); an order of magnitude of 10^{10} viruses is the accepted one.

that the virus was cleared very rapidly from chronically infected patients, whereas the maintenance of its stable levels required an enormous rate of production. On the other hand, despite the rapid response to therapy, the high mutation rate (see chapter 2) along with the production levels implied that, during HIV-1 replication, every single possible mutation would be present (Coffin 1996; Perelson et al. 1997). As it became soon clear, the HIV-1 quasispecies would become easily resistant to any single drug. The rapid production and clearance of HIV-1 revealed also that, whatever was taking place during the latency phase, it was likely to involve a highly dynamical phenomenon.

In the next two sections we will take a step further in order to provide two alternative (not necessarily exclusive) explanations of the latency phase. The first considers the impact of spatial dynamics in creating a delay in the expansion of the virus due to the constraints created by limited growth within lymph nodes. The second instead explicitly introduces genetic variability and thus evolutionary dynamics.

4.4 Spatial Dynamics of HIV-1

A key component of the organization of immune responses within our bodies is given by a network of connected elements known as lymph nodes, where $CD4^+$ T cells are produced (Janeway et al. 2001). Lymph nodes have a complex structure and are also known to be a reservoir of the majority of HIV-1 particles (Kelly and Taiwo 2015; Lorenzo-Redondo et al. 2016). The model assumes that interactions between $CD4^+$ T cells and HIV-1 take place on a space that can be assimilated to a square lattice (Zorzano dos Santos and Coutinho 2001). This is partially justified by the spatial structure of lymph nodes, which are somewhat similar to a mesh (Willard-Mack 2006).

Cellular automata models have been used to model immune responses associated to viral infections (Stauffer and Pandey

1992; Papa and Stallis 1996). The model considered here assumes a square lattice Ω with $N = L \times L$ sites, where each site $S_i \in \Omega$ can be in one among four possible states from a set

$$\Sigma = \{H, I_1, I_2, D\}, \tag{4.18}$$

where H=healthy site, I_1=infected cell capable of spreading viruses, $I_2 =$ aging infected cell, susceptible to the immune attack, and D stands for dead (empty) sites. Initially, the lattice is filled with $N(1 - p_{\mathrm{HIV}})$ H cells and only a very small fraction of Np_{HIV} occupied by $(S_k = I_1)$ infected cells. Afterward, the following set of rules applies:

1. Infection: $H \rightarrow I_1$ for a site S_i occurs if at least one neighbor S_j of this site cell is infected (i.e. $S_j = I_1$). If this condition is not met, infection can also occur if at least R neighbors are of I_2 class.
2. Decay of infected cells: A transition $I_1 \rightarrow I_2$ occurs after τ time steps. Here τ would represent the time scale for developing a response aimed at killing infected cells.
3. Death and cell removal: a transition $I_2 \rightarrow D$ occurs with probability 1.
4. Colonization: expansion into empty sites $D \rightarrow H$ occurs with probability $p_r(1 - \epsilon)$. Otherwise, we have $D \rightarrow I_1$ with probability $p_r \epsilon$. Here ϵ is the probability of spontaneous infection.

In figure 4.5 we display the outcome of this model for a given set of parameters;[5] specifically a square $L \times L$ lattice with $L = 700$ was used. Here the time series includes healthy cells (open squares), infected cells (full circles), and dead cells (open triangles), respectively. We can see the close similarities between

---

[5]Most parameters have been calibrated using available information to provide proper orders of magnitude.

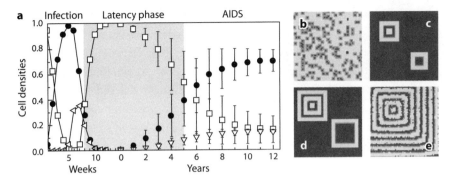

Figure 4.5. Time evolution of the cellular automaton model for HIV-1 immune system interactions on a lattice. In (a) the populations of infected (filled circles), healthy (open squares), and dead (triangles) cells are displayed, respectively. In (b-e) four snapshots of one of the runs are shown. Different gray levels correspond to different states. Here we can see how propagation occurs in waves that create an effective delay in achieving a global infection level.

this model results and those reported from blood samples of HIV-1-infected patients. In particular, the number of infected cells I_1 clearly reflects a lagged response that is consistent with the latency phase. The results of these simulations have been averaged over 500 runs (this is why error bars are shown) using different initial random conditions.

After a rapid initial phase of infection with high numbers of infected cells (whose numbers provide a surrogate of the viremia) a rapid decline occurs followed by a latency phase and, later on, the expansion of infection consistent with the AIDS dynamical pattern. The characteristic time introduced by τ leads to the formation of waves with length $\tau + 1$. After an initial infection starting from a random subset of sites, these seeds originate a rapid spread and most sites will get infected (figure 4.5b, brighter area) after about five weeks in the simulation. After $2\tau + 1$ steps, the number of infected cells strongly decays and is followed by rare, new infections on a lattice dominated by H cells (figure 4.5c). These new seeds initiate a structured spatial process involving propagating fronts (figure 4.5 d-c). Because of these slow-expanding waves, infection requires now a much longer time to spread and cover the lattice, thus causing a marked latency. Other spatial models have been developed providing consistent results (Perelson and Weisbuch 1997; Bernaschi and Castiglione 2002; Shi et al. 2008; Perelson and Shuai 2010).

4.5 Antigenic Diversity Thresholds and AIDS

The cellular automata model includes the spatial structure of lymph nodes (in an oversimplified fashion) and provides a potential explanation of the latency phase. In this case, latency is largely due to the spatial constraints that limit the propagation of infections through the system. However, despite the qualitative matching, some important variables and rules have been ignored. One in particular is the high potential for mutation displayed by RNA viruses (including HIV-1), which is largely responsible for adapting and evading immune responses. On the other hand, the mirror image of this fast change is provided by the immune system responses and its intrinsic diversity. These are features that can be observed and measured, and it was soon realized that HIV-1 was capable of developing great variability both among patients as well as over time within the same patient (Cichutek et al. 1992; Sanjuán et al. 2004).

Many models of HIV-1-immune system interactions have been developed (Arnaout et al. 2000; Bittner et al. 1997; Callaway et al. 1999; Nowak et al. 1991, 1996; Wikramaratna et al. 2015; Nowak and Bangham 1996; Wodarz and Nowak 1999, 2002) from different perspectives and involving several levels of complexity. A common feature of all of them is the presence of a diverse set of viral strains and a specific cellular response mediated by strain-specific $CD4^+$ T cells. During the asymptomatic phase of infection (figure 4.2), error-prone replication of HIV-1 generates increasing numbers of antigenic variants (i.e., viruses that differ in their coat proteins and thus present different epitopes to the $CD4^+$ T cells). During the early period of infection, the immune system is capable of coping with this increasing variability and regulates the growth of the HIV-1 population. However, there is a threshold value in antigenic variability, and this is the key component of the model, above which the immune system cannot exert effective control and the viral population induces the collapse of the $CD4^+$ T lymphocyte population. Therefore, the model asumes that HIV-1 antigenic diversity is the cause of AIDS and not its consequence (Nowak et al. 1991). In the following paragraphs we will present the fundamental properties of this model.

Instead of lumping together all viral strains into a single phenotypic class, we introduce viral diversity by modeling the underlying fitness landscape of HIV-1: different viral strains correspond to different genomes, each one leading to a different phenotype. Empirical evidence suggests that the underlying HIV-1 fitness landscape is quite complex, characterized by regions of high ruggedness (that depend on the type of epistatic interactions among mutations; see chapter 3) interrupted by flat regions where mutations are effectively neutral (Bonhoeffer et al. 2004; Hinkly et al. 2011; Kouyos et al. 2012). Interestingly, these properties of the fitness landscape do not strongly depend on whether viral fitness is measured in very permissive conditions or in harsh ones owing to the presence of antiviral drugs (Kouyos et al. 2012). Here we will limit ourselves to a very simple interaction scheme where specific HIV-1 strains interact only with some specific strains of $CD4^+$ T cells capable of clearing the virus but also being damaged by its infection. The specific nature of the response is introduced in a minimal model with $2n$

equations (Nowak 1992) defined as:

$$\frac{dv_i}{dt} = v_i(r - px_i) \tag{4.19}$$

$$\frac{dx_i}{dt} = kv_i - uvx_i. \tag{4.20}$$

Here x_i and v_i indicate populations asssociated to the ith virus strain and its corresponding CD4$^+$ T-cell partner. For simplicity (this is an obvious oversimplification) we assume that all viruses replicate at the same rate r and the immune response given by the term px_iv_i is also homogeneous. Similarly, the growth and death terms in the second equation use identical parameters for all strains.

Summing over all the previous equations, we obtain for the total viral population the expression

$$\frac{dv}{dt} = v\left(r - p\sum_{i=1}^{n} x_i\frac{v_i}{v}\right) \tag{4.21}$$

and for the corresponding immune cell response,

$$\frac{dx}{dt} = kv - uvx, \tag{4.22}$$

where we have used $x = \sum_i x_i$. The virus population will be under the control of the immune response provided that

$$\frac{r}{p} < \sum_{i=1}^{n} x_i\frac{v_i}{v}. \tag{4.23}$$

From the equation for x_i, we can see that the immune response leads to a stationary state

$$x_i = \frac{kv_i}{uv}. \tag{4.24}$$

Using this result and applying it to the total viral population, we obtain an equation for the overall virus population that depends on the actual diversity:

$$\frac{dv}{dt} = v\left(r - p\frac{k}{u}D\right), \tag{4.25}$$

where D stands for Simpson's diversity index (Magurran 1988), namely

$$D = \sum_{i=1}^{n}\left(\frac{v_i}{v}\right)^2, \tag{4.26}$$

and is used in theoretical ecology as an inverse measure of diversity. For a homogeneous population, with all individuals belonging to the same species (strain), we have $D = 1$ (maximum), whereas in a completely heterogeneous system where all species are equally represented, we have $D = 1/n$.

This is a very interesting result, which connects the deterministic model to a statistical index related to ecological diversity. In our problem, this actually measures antigenic diversity as defined by the phenotypic differential traits displayed by the viral population. From the previous equation for the viral load, it is clear that it will decline or increase depending on the diversity D compared with the critical value

$$D_c = \frac{ru}{pk}. \tag{4.27}$$

If $D < D_c$ we will have $dv/dt > 0$, and *the virus escapes to the immune system's control.* If we make the additional approximation that $D = 1/n$ (uniform viral population), the previous critical condition can be expressed in terms of the total number of different strains present, i.e., the condition for viral escape is:

$$n > n_c = \frac{pk}{ru}. \tag{4.28}$$

This model thus provides an elegant explanation of what is taking place throughout the latency phase. It also allows us to understand it in terms of a virus-immune system arms race and makes a well-defined prediction. The model suggests (as seems to be the case) that the long and apparently calm interval of low viremia and slow decay of $CD4^+$ T cells hides a rapid turnover of constantly emerging viral escape mutants while progressive damage to immune responses takes place. This would explain the apparent dormancy of the virus: in reality there is a constant battle between host defenses and the new variants being created through mutation. This battle keeps the viral load at low levels, but also involves a high cost due to the constant turnover of $CD4^+$ T cells.

The prediction is that, as antigenic diversity grows over time (more and more mutants are likely to persist), the potential for response decreases until collapse takes place, with a rapid increase in viral diversity, measured in both numbers of present strains and Simpson's index. The outcome of the model is exemplified in figure 4.6. The model starts with a single strain and new strains are added within a time interval dt with a probability $Pv(t)dt$, thus proportionally to the viral load (here $P = 0.1$). Figure 4.6a displays the total population of viruses $v = \sum_i v_i(t)$, which has a very slow progression and a long transition with low levels followed by exponential growth at the equivalent of the in silico AIDS phase. Figure 4.6b instead shows the detailed population growth and decay of different variants of the virus, which are more and more frequent as time proceeds. Finally, figure 4.6c shows the antigenic diversity, measured as both the total number of observed strains (continuous line) and the Simpson's diversity index. By using a more detailed implementation of the HIV-1 quasispecies, it is possible to introduce a mutation matrix connecting different viral strains, as shown in figure 4.7 for a given matrix and a given set of initial conditions. As displayed on the time series of HIV-1 strains, the three phases are easily identified.

The model so far discussed can be extended to accommodate additional, relevant components of the immune response. In particular, one missing element in our previous assumptions is

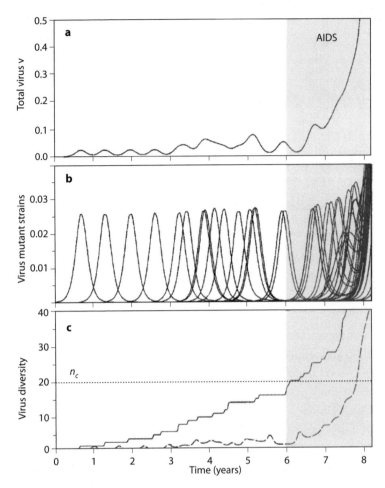

Figure 4.6. A numerical simulation of the simplest model that includes
antigenic variation in the HIV-1 population, the stimulation of specific
immune responses, and the impairment of the immune function. The
parameters used are $r = 1$, $p = 20$, $u = 1$, and $k = 1$. These parameters
predict a critical value $n_c = 20$ (the gray areas indicate $n > n_c$). Here
we have: (a) Total virus population size (virus density in arbitrary units);
(b) abundance (density) of the individual HIV-1 variants, and (c) the total
number of strains (continuous line) and the inverse of the Simpson index
(broken line) as a measure for population diversity.

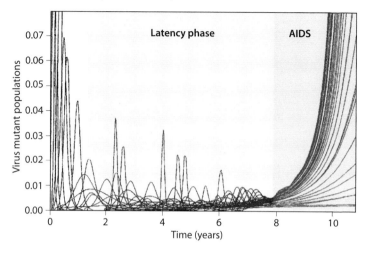

Figure 4.7. A numerical simulation of a quasispecies model (adapted from Nowak (1992)). The model incorporates the mutation matrix connecting different HIV-1 strains.

the consideration of nonspecific responses (see Nowak and May 2000 and references therein). The model considered a one-to-one $(x_i \leftrightarrow v_i)$ matching between viral and cell strains, and thus lacks cross-reactive responses. However, such an assumption ignores a well-known observation: some parts of the virus genome (and thus of its associated proteins) are highly conserved and less prone to mutation. In terms of the immune system, that means that these are efficient targets that can allow control of the infection. This could explain the more rich spectrum of possible progression pathways reported from some HIV-1-infected individuals. In this context, along with the previous three-phase development of AIDS, it has also been observed that some patients rapidly develop the disease with no previous delay whereas some do not display the symptoms even many years after the infection. Can we explain these three outcomes with a single model?

Cross-reaction can be introduced by relaxing the assumption that immune responses are only specific. To illustrate how this can be done, let us expand the previous model as follows (Nowak and May 2000):

$$\frac{dv_i}{dt} = v_i(r - px_i - qz) \qquad (4.29)$$

$$\frac{dx_i}{dt} = kv_i - uvx_i \qquad (4.30)$$

$$\frac{dz}{dt} = kv - bz, \qquad (4.31)$$

where z incorporates the cross-reactive immune response (as we can see, if $q = 0$ we have the previous model). As defined by the third equation, this nonspecific response is also stimulated by the presence of viruses and each viral strain can be removed at a rate qzv_i. Despite the apparently small change introduced, the new model displays now three different classes of behavior, reported from the study of HIV-1 and other lentiviruses. Very shortly, these phases are:

1. Lack of latency phase. It occurs for $ru > kq + cp$, and the virus here rapidly expands with no delayed growth.
2. Chronic infection with no disease. This phase occurs when $kq > ru$. In this case, viruses get established and are eventually controlled and kept at low abundance.
3. Chronic infection leading to disease. This is observed within the interval of parameters $kq + cp > ru > kq$ and corresponds to the AIDS scenario discussed above with a latency phase.

The results from this model illustrate how the combination of specific and nonspecific responses provide a diverse (but finite) set of possibilities concerning the outcome of infections in lentiviruses. More generally, they give us a powerful picture of the evolutionary dynamics responsible for the asymptomatic phase and its meaning in terms of a virus-immune system arms race. In the next section we will explore this idea in a different context, illustrating the importance of evolutionary dynamics in shaping the architecture of the interaction webs among viruses and their hosts.

The models described in the previous sections need to be seen as useful ways of reducing the overwhelming complexity of virus-host exchanges in order to achieve some understanding of the events taking place at a given scale. Other modeling approaches are required to analyze the treatment of AIDS patients, which needs proper statistical and immunological models (Sloot et al. 2005; 2006) to generate patient-specific medical simulations (Sadiq et al. 2008a). Similarly, the design of antiretroviral inhibitors call for molecular simulation tools as close as possible to the physicochemical events (Sadiq et al. 2008b). But they must not be seen as better alternatives to the simple mathematical models discussed above. Instead, as with other complex systems, multiple scales are at work and different goals are associated to the modeling paths taken in each case.

To close this section, we think it is worth mentioning the existence of an alternative recent model that gives a more relevant role to antibody responses in the control of viremia, and suggests that escape from or progressive loss of quality of $CD8^+$ T-cell responses are associated with an acceleration of disease progression, the underlying cause of the breakdown of virus control is the loss of antibody induction due to depletion of $CD4^+$ T cells (Wikramaratna et al. 2015). This more complex model does better match with genomic data that indicate repeated selective sweeps that result in purging antigenic diversity (Walker and Korber 2001; Frost et al. 2005; Wibmer et al. 2013).

4.6 Viral Symbiosis

So far the general view of viruses emerging from previous chapters offers a picture of these entities dominated by their parasitic

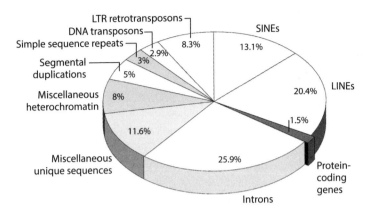

Figure 4.8. Frequency of different components of the human genome. There is only a small percentage of the total genome that seems to play a standard functional role, namely genes transcribed into proteins (just 1.5%), but a very high frequency of diverse genetic elements including transposons and endogenous retroviruses.

nature. We have however already mentioned that this is an incomplete perspective and that the more we know about viruses the more we can appreciate a broader picture. We have discussed in chapter 2 the special nature of giant viruses, escaping from the simplified view of viruses as pieces of software infecting the cellular hardware. But there is more than that. In fact, viruses can have beneficial effects on their hosts, creating a symbiotic relationship (Roossinck 2011; Ryan 2009). And this might be far from anecdotic. Many examples of viruses that provide functional benefits to their hosts are known to create a mutualistic tie (Roossinck 2011).

Strong evidence of a special relation between viruses and their host organisms (and primates in particular) is provided by the widespread presence of so-called endogenous retroviruses (ERVs), which account for 8% of the total DNA content of the human genome (Sverdov 2000; Belshaw et al. 2004). What is the evolutionary significance of ERVs? One sound possibility, supported by well-documented studies (Tarlinton et al. 2006) of Australian koalas, indicates that endogenization of ERVs has taken place, protecting their carriers against other lethal pathogens (Denner and Young 2013).

More generally, we face a completely different scenario where viruses infect the germ line, which allows long-term persistence of viruses (Villareal 1997; Katzourakis et al. 2005). As pointed out by Ryan (2009), a new evolutionary dynamics enters the scene, allowing for novelties, and for selection to operate at the level of host-virus interaction. This includes the possibility of an evolving symbiotic relationship. Perhaps the best illustration of this potential role in major evolutionary leaps is given by the evolution of placental mammals (Harris 1991; Haig 2012).

The development of the placenta[6] requires the presence of ERVs. Specifically, the proper development of this multifunctional organ proceeds through a process of cell fusion that takes place between cells belonging to different tissues. Syncytin, a key protein required for the fusion process, is provided by an endogenous retrovirus (Villareal 1997; Haris 1998; Mi et al. 2000; Rote et al. 2004). As pointed out by David Haig, the origin of the placenta created a new niche that provided new opportunities for ERVs to pass from mother to offspring. The resulting arms race between host defenses and evasion mechanisms led to a final, tight mutualistic relationship (Haig 2012).

Other remarkable examples include a long-term coevolution between a large class of viruses (the polydnaviruses) and their host wasps (Villareal 1997; Roossinck 2011). These wasps are parasitoids: they are different from standard predators, since they lay their eggs inside the larvae of their prey species, which develop inside the body of the living larvae by eating them from inside. The normal outcome of this should be an immune response

---

[6]The virus-driven evolution of the placenta took place independently at least six times (Haig 2012 and references therein).

capable of encapsulating the injected eggs and inhibiting egg development. However, the endogenous virus carried by the wasp egg suppresses this response. The coevolutionary ties are very strong and some authors have questioned how appropriate it is to consider the polydnavirus as a real virus.[7]

From a more general perspective, viruses have played major roles in promoting speciation and host innovation (Villareal 1997), even driving the host to increase in genome complexity. These arms races have shaped ecological webs and have also affected our own history. In the next two chapters, we explore how viruses propagate through populations and how social and transportation networks might affect epidemic outbreaks.

---

[7]As an example of the complex relationship between both partners, let us mention that the genes required for viral replication and packaging have moved to the wasp genome.

5

EPIDEMICS

5.1 Outbreak

The history of humankind is often written under the names of great heroes and powerful emperors (McNeill 1976; Diamond 1997; Watts 1997; Clarck 2010). Large armies of soldiers, we are told, were able to conquer a world where rational human decisions were made and the right strategies were adopted. This is often the accepted story that explains unique events, such as the surprising defeat of the Aztecs by the Spanish conquerors. Hernán Cortés (figure 5.1, left) became a national hero after he defeated the powerful Aztec empire despite the size of the latter. Heading no more than six hundred men, Cortés confronted a whole civilization with 500 warriors ready to fight each Spaniard. Although horses and guns initially had a profound impression on the natives, they also soon realized that the conquistadores were weaker and were ready for retaliation. Despite their larger armies and knowledge of the territory, the Aztecs were defeated by the small army of invaders. Cortés won the battles and the old Aztec gods and emperors were replaced by new ones.

There is little doubt that Cortés, like other conquistadores before and after him, was bold and took risks, but a major player was also involved, unnoticed and not to be mentioned in

 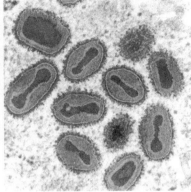

Figure 5.1. Viruses shaping history. Hernán Cortés became famous as the conqueror of the Aztec empire despite his vastly inferior army. However, the Spaniard came to the New World carrying a deadly companion: the smallpox virus, which rapidly propagated within the native population, killing or seriously debilitating much of it.

most tales to be written. Along with those pale men, an invisible warrior was reaching the other side of the world, where it would spread among a human population that was defenseless: the *Smallpox virus* (figure 5.1, right). Without the immunization that was known to the colonizers, smallpox found an entirely empty niche formed by the whole population of Americans unable to understand that there was no magic or high power in the resistance of the Spaniards to the disease. None of the soldiers died from the pestilence, whereas a majority of their people, once infected, either died (in a terrible way) or became seriously ill. This deadly pathogen was well known in the Old World and its mark has been traced back in history to ancient Egypt, but it was alien to the New World.[1]

---

[1] This observation can actually rule out the suggestion that Chinese explorers and merchants had been in contact with the Old World long before Columbus arrival. If that were the case, a pandemic should have been triggered, leaving a detectable mark.

It has been argued that epidemics have played a major role in human history (McNeill 1976; Diamond 1997), and indeed this seems to be the case. The encounter between the Spaniards and the native Americans is just one example of the predictable outcome of colonization of distant parts of the world by foreign (mostly European) explorers. The arrival of the British in Australia triggered a deadly wave of smallpox infections, with the estimated death of half of the indigenous Australians. The same virus wiped out the population of Easter Island, and measles eliminated a third of the inhabitants of Fiji. Many of these events, which became pandemics (i.e., large-scale, often worldwide events) involved "new" viruses, not known before their spread claimed vast numbers of victims. This is the case of HIV-1 (Holmes 2001; Rambaut et al. 2001; Hemelaar 2012; Faria et al. 2014), which became a major threat after its epidemic spread from Africa in the late 1970s, though its origin has been dated to the 1920s in central Africa (Faria et al. 2014). The death toll of these outbreaks is summarized in figure 5.2. Some of these have been devastating, in some cases damaging the economy and society of whole countries, such as South Africa. Some of them are not much remembered because they were brought under control long before their spread was effective (this is, e.g., the case of the coronaviruses causing the severe acute respiratory and the Middle East respiratory syndromes, SARS and MERS, respectively).

New threats have been emerging as human populations explode and our pressure on ecosystems crosses sustainability thresholds (see chapter 6). But, as will be shown below, there is a bright side to the dynamical behavior of many of these epidemic events that makes possible to achieve the impossible: the complete eradication of some of them. A dramatic illustration is given by the complete, worldwide elimination of smallpox, despite the hundreds of millions of deaths that it caused. In December 1979, its eradication was officially declared (Ellner 1998; Lane 2006).

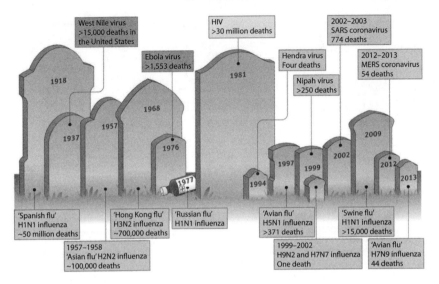

Figure 5.2. Virus outbreaks have devastated human populations through history. Some usually harmless outbreaks, such as those caused by the ordinary influenza, have sometimes killed millions. The HIV-1 virus also produced a major outbreak that took the life of millions since its appearance in the 1970s. New outbreaks of these pathogens can occur again as new variants and ecological conditions emerge. Adapted from Bean et al. (2013).

To understand how to deal with epidemics and eradicate them, we need first to understand how they spread. In figure 5.3a we show the fraction of months without measles (MWM) in populations of different size N. The plot reveals a remarkable nonlinear response, suggesting that, once a critical population size is reached, epidemic spreading becomes inevitable. We also display (figure 5.3b-c) the dynamics of reported cases of measles (a) and deaths caused by it (b) in the US from the 1950s. We can clearly appreciate that a shift occurs at the beginning of the 1960s, when the measles vaccination started to be used. Once vaccines became operative, the potential for further outbreaks diminished to a residual level. This pattern is common to most other vaccine-

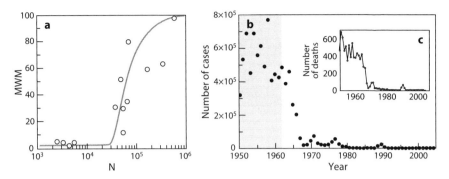

Figure 5.3. Nonlinear dynamics of epidemics. In (a) we display the fraction of MWM displayed by different human populations. Some pandemic diseases have been virtually eliminated once vaccination has been introduced. The gray line is a suggested fit that is consistent with the theory presented here. In (b) we show the number of measles cases in the US whereas (c) displays the number of deaths caused by this disease. The end of the gray area marks the initiation of the measles vaccination.

preventable diseases and suggests that responses to vaccines are highly nonlinear.

The likelihood of a virus (or any pathogen) spreading within a given population is tied to a number of biological traits but it also relates to other variables, including the behavior of individuals: how they live, how they crowd in cities or hospitals, and how they travel. Cultural practices can also play a role, as illustrated, for example, in the mourning of the dead in some parts of Africa involving close contact with the corpses. Since the EBOV remains active for days after the death of its human host, corpses are an important factor in promoting its epidemic spreading. And technology has also changed the landscape of pandemics. Two hundred years ago, the world was pretty much flat and local: transportation required extended periods of time and there was no way to escape from the constraint of slowly moving carriages or horses to cover distances.

The rise of industrial civilization and the increasing dominance of cities clearly influenced the impact of epidemic spreading. Once horse carts were replaced by fast, long-distance transportation, the potential for pathogens to spread and eventually affect all human beings became inevitable. The invention of commercial flights changed everything. Just a few decades after the first short-lived experiment performed by the Wright brothers in 1903, hundreds of humans were using aerial transportation. How can these factors be incorporated in a modeling framework? In this chapter we present some key results that allow us to explain the nonlinearities and breakpoints that characterize epidemics, including their rise and fall.

5.2 SIS Model

Epidemic modeling has been a very active area of research for several decades (Anderson and May 1992; Keeling and Rohani 2007), with an enormous practical impact on our understanding of the dynamics of infectious diseases. One of the simplest models considers just two subpopulations, namely *infective* (I) and *susceptible* (S) individuals. The first are infected and carry the pathogen with them while the second are pathogen-free and can become infected. The states of these individuals can change due to two basic processes: recovery ($I \rightarrow S$) and infection ($S \rightarrow I$), which occur with some fixed rates (that can be determined from epidemiological data). While the recovery process is independent of the state of other individuals, the infection process is not: the more individuals are infected, the larger the probability of infection.

The basic logic of the model, which contains the rules associated with transitions among alternative states, is shown in figure 5.4a. The simplicity of this model can be appreciated by comparing it with other more accurate ones, including the SIR (for susceptible-infected-recovered) model (figure 5.4b). Here an

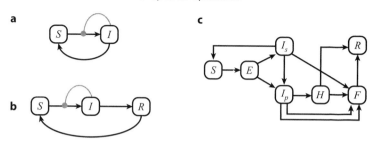

Figure 5.4. Transition graphs for different scenarios of epidemic propagation. Here we include (a) the SIS model, (b) the SIR model, (c) the SEIR model, and an example of a more complex transition diagram (d) associated with the Ebola virus. Susceptible (S), Exposed (E), Infectious (I), Hospitalized (H), and Funeral (F) indicate transmission from handling a diseased patient's body, in contrast to Recovered/Removed (R).

additional compartment has been included (R), incorporating an additional population of recovered individuals that are temporarily immune to infection. In other cases, such as the epidemic spreading of EBOV, a more complex set of transitions needs to be added (figure 5.4c), including two additional classes of infected individuals as well as hospitalized and unburied corpses.

In this section a detailed analysis of the SIS model is introduced. This is a toy model, and some basic assumptions are required. One is that $I + S = N$, which means that, at some scale, the total population remains constant. Secondly, there is no heterogeneity and thus all interactions between individuals are weighted with exactly the same parameter values. In a well-mixed system, the rules outlined above can be described as two reactions associated to infection and recovery, given by

$$I + S \xrightarrow{\mu} 2I \tag{5.1}$$

$$I \xrightarrow{\alpha} S, \tag{5.2}$$

and it is easy to show that the equations describing our system are

$$\frac{dI}{dt} = \mu IS - \alpha I \qquad \frac{dS}{dt} = -\mu IS + \alpha I, \qquad (5.3)$$

which, as we can see, gives $dI/dt = -dS/dt$; and since the total population $I + S$ is conserved, using a normalized density of infected individuals, $\rho = I/N$, we have

$$\frac{d\rho}{dt} = \mu\rho(1 - \rho) - \alpha\rho, \qquad (5.4)$$

consistently with our previous result. We have a one-dimensional model that can be solved by integrating the previous equation. This equation can actually be rewritten as a logistic-like form:[2]

$$\frac{d\rho}{dt} = \mu\rho(\rho^* - \rho), \qquad (5.5)$$

where $\rho^* = 1 - \alpha/\mu$. If ρ_0 is the initial population of infected individuals, we obtain:

$$\rho(t) = \frac{\rho^*}{1 + \left(\frac{\rho^*}{\rho_0} - 1\right) e^{-(\mu-\alpha)t}}. \qquad (5.6)$$

The previous equation provides two interesting results. One is the steady state reached for long times, i.e., for the limit $\rho_\infty = \lim_{t\to\infty} \rho(t)$, we have two possible solutions, namely $\rho_\infty = \rho^*$ when $\mu > \alpha$ and $\rho_\infty = 0$ when $\mu < \alpha$. The first point involves a stable epidemic event that would infect a fraction $(1 - \alpha/\mu)$ of individuals of the population, whereas the second represents the extinction of the virus. The critical point $\mu_c = \alpha$ separates the subcritical phase, where the epidemic dies out, from the supercritical phase, where the epidemic is self-maintained.

These two possible states are associated with two qualitatively different phases, as it occurred with the numerical experiment

---

[2]In population dynamics, one of the simplest equations describing growth of a population N under limited resources is the *logistic equation*, defined as $dN/dt = rN(1 - N/K)$, where K is the carrying capacity.

described above using an SIS model on a lattice. We can compare the predicted infected population with the previous result involving a spatial patterning. We have also run the previous model, but in this case once we choose an empty site and then choose a random site (not a nearest neighbor) that is infected, the first can be infected too with a probability μ. The exact location of the transition point has moved and the shape of the curve on the right-hand side is slightly different, but the phenomenon itself remains preserved.[3]

A second implication of the previous solution $\rho(t)$ is that, at the supercritical phase, the initial growth of the epidemics is exponential. This can be shown by assuming that the current relative frequency of infected individuals is very small, i.e., $\rho \ll 1$. In this case, it is possible to make the approximation $1 - \rho \approx 1$, and thus the equation for epidemic spreading now becomes

$$\left(\frac{d\rho}{dt}\right)_{\rho \ll 1} = (\mu - \alpha)\rho, \qquad (5.7)$$

which is easily solved and leads to an exponential solution:

$$\rho(t) = \rho(0)e^{(\mu - \alpha)t}. \qquad (5.8)$$

This approximated solution holds in early phases of an outbreak; and we will observe a rapid propagation provided that $\mu > \alpha$ (otherwise the epidemic dies out). As defined above, it is easy to see that the previous equation can be also written as

$$\rho(t) = \rho(0)e^{\alpha(R_0 - 1)t}, \qquad (5.9)$$

where R_0 is the so-called *basic reproductive number* R_0, defined here as

$$R_0 = \frac{\mu}{\alpha}. \qquad (5.10)$$

---

[3]The use of space makes propagation more difficult, since infection is limited to a small number of neighbors instead of potentially affecting *any* individual.

Table 5.1. Examples of viral pathogens involving different forms of transmission between individuals and their characteristic basic reproductive numbers R_0.

Disease	Transmission Route	R_0
EBOV (2014 outbreak)	Body fluids	1.5–2.5
HCV	Blood-to-blood contact	1.2–1.7
HIV-1	Sexual contact	2–5
IAV H1N1 (1918 pandemic)	Airbone droplet	2–3
Measles	Airbone	12–18
Mumps	Airbone droplet	4–7
Polio	Fecal-oral route	5–7
Rubella	Airbone droplet	5–7
SARS coronavirus	Airbone droplet	2–5
Smallpox	Saliva	6–7

This result captures a fundamental message: propagation occurs provided that R_0 is larger than 1.[4] This parameter is usually defined as the expected number of secondary infections produced by a single (typical) infection in a completely susceptible population. Table 5.1 shows the R_0 values estimated for some well-known viral diseases. Values widely range from 1.2 (for genotype 6 of HCV) to 18 (for the measles virus) and are clearly influenced by the route of transmission. Epidemic spreading will occur if $R_0 > 1$ and die out instead for $R_0 < 1$. It is interesting to note that R_0 involves several components, including the infectivity of the pathogen μ but also the population size N.

An important message needs to be taken from the early growth equation: if the pathogen has an $R_0 > 1$, the outbreak will result in exponential growth, meaning that unless the process is controlled at early stages, it can easily escape out of control. Such a situation (as discussed below) is exacerbated by the modern

[4]If we normalize our population to $I + S = N$ we obtain $R_0 = \mu N/\alpha$, thus including the population size N as a key component. This is actually consistent with the transition found in figure 5.3a, where a critical $N_c \approx 3 \times 10^4$ seems to be at work.

scenario of widespread and fast human mobility patterns, which strongly favor the spread of the pathogen. However, knowing the exact value of R_0 is also the key for limiting or even stopping the epidemic outbreak. It is easy to show that, when a fraction r of individuals in the population has been vaccinated or simply isolated from others, the basic reproductive number gets corrected to $R_0 = 1/(1 - r)$. That means that proper action on a fraction of the total population can trigger an epidemic to die out. It can be shown, for example, that for the EBOV case with $R_0 = 1.5$, just a third ($r \approx 1/3$) of the individuals need to be treated. This is in fact the basis of vaccination, which does not necessarily require a whole-population treatment. In some cases, however, high R_0 numbers require vaccination of all involved agents.

5.3 SIS Model in Space and Graphs

The previous equations defining the SIS model dynamics can be extended in many ways to incorporate more elements of realism. On the one hand, individuals do not interact in a fully mixed fashion: to some extent, infection is a process that requires locality as we are more likely to interact with some subsets of individuals than with others that live beyond some distance. On the other hand, the use of transportation systems also has a major impact on the patterns of infection. Moreover, models described by differential equations miss an important ingredient: the stochastic nature of epidemic spreading. In order to address these missing parts, it is convenient to take a new look at the SIS model from a microscopic perspective.

Consider a given system composed by N individuals interacting through these SIS rules. For simplicity, assume that each individual in this system occupies a node in a network. Interactions here are described by edges. The resulting graph can, for example, be random (figure 5.5a), with nodes having a probability p of being connected to each other. This is the so-called Erdös-Rényi random graph. In general, if the population can be represented as a network Γ where vertices v_i (with $i = 1, \ldots, N$) are individuals and edges (links) between them indicate potential interactions, a given individual can have k_i (with $1 \leq k_i \leq N$) edges. This quantity is known as the *degree* of the v_i agent (Newman 2010) and the *average degree* of the graph is given by

$$\langle k \rangle = \frac{1}{N} \sum_{i=1}^{N} k_i, \tag{5.11}$$

and provides a statistical measure of the network connectivity. Moreover, this graph has an associated distribution of connections, or *degree distribution*, $P(k)$, that follows a binomial shape,

$$P(k_i = k) = \binom{N-1}{k} p^k (1-p)^{N-k-1}. \tag{5.12}$$

Here $P(k)$ is a homogeneous distribution, with a well-defined average degree[5] $\langle k \rangle = p(N-1)$. An example of a random graph of this type (if we ignore the location of nodes) is shown in figure 5.5a.

A lattice (figure 5.5b) is another limit case of a graph where each node has exactly the same number of neighbors, i.e., $P(k) = \delta_{qk}$, for all vertices.[6] An intermediate situation is given by figure 5.5c, where interactions are random but spatially limited to nearest sites. The rules of the SIS model described above can be easily mapped into our networks. In figure 5.6 we show how this works. Consider a given noninfected site (open circle) with just three neighbors. Suppose that this node can be infected with probability μ if one of these neighbors is

---

[5]More generally, the average degree can be obtained from $P(k)$ by means of the integral $\langle k \rangle = \int_{k_0}^{K} P(k)dk$, with k_0 and K indicating the limit values of k.

[6]The distribution δ_{qk} is the so-called Dirac's delta function, defined as $\delta_{qk} = 1$ when $k = q$ and zero otherwise.

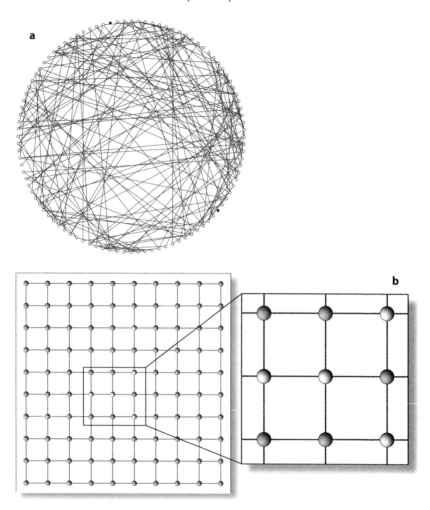

Figure 5.5. Networks of interactions. Two different examples of networks are shown here, where nodes (empty circles) represent individuals in a population and their links indicate the presence of (potential) interactions. The examples include (a) a random graph where every two elements (randomly scattered) are connected to each other with some probability and (b) a square lattice, with each individual having exactly $q = 4$ neighbors (zoomed area).

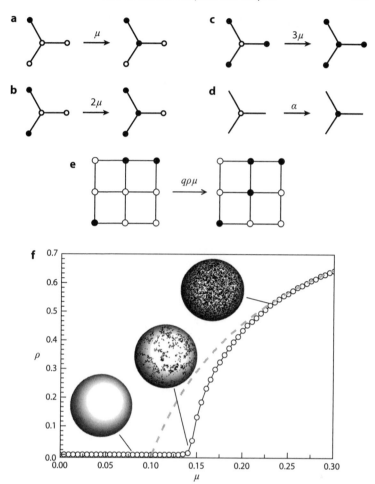

Figure 5.6. Phase transition in the SIS model. The rules are indicated in (a-d) for a system where each element is connected to just three neighbors, here $\bullet \equiv I$ and $\circ \equiv S$, respectively. For a given μ, infection (a-c) will be more likely as the fraction ρ increases. Additionally, recovery (d) occurs with a probability α. In an arbitrary case, such as a lattice (e) with q nearest sites, the probability of infection will grow proportionally to $q\rho$. In (f) we display ρ against μ using a square lattice Ω with $L = 100$ and $\alpha = 0.1$. A marked transition occurs at $\mu_c(\Omega) \approx 0.15$. Three spatial snapshots are also indicated (inset).

infected (figure 5.6a). If more than one is infected, for small μ this probability grows linearly[7] (figures 5.6b-c). The recovery rule instead is independent of the states of the neighbors (figure 5.6d).

The SIS model reactions defined above can be easily expanded to arbitrary networks, including regular lattices (Newman 2010). The previous algorithm can be easily extrapolated to other networks and lattices. In figure 5.6e we indicate this for a square lattice having a fraction ρ of infected nodes and a given number q of nearest sites. For this particular case, where epidemic spreading is confined to nearest sites, a stochastic cellular automaton is defined on an $L \times L$ lattice $\Omega = \{r = (i, j) | 1 \leq i, j, \leq L\}$. The state of each site $\sigma_t^t \in \{0, 1\}$ at a given step t indicates the presence of infected (1) or susceptible (0) individuals. The simulation starts from a random initial condition with a fraction of infected sites $\sigma_x^t = 1$ and the rest set to zero (susceptible). The rules of the simplest version of the model (see Solé and Bascompte 2006 for details) are:

1. Choose a site $r \in \Omega$ at random.
2. If $\sigma_r = 1$, a transition to $\sigma_r = 0$ occurs with probability α.
3. If $\sigma_r = 0$, choose a random neighbor σ_u. If $\sigma_u = 1$, an infection event will occur with probability μ.
4. Repeat (1)

In figure 5.6f we show the results of a numerical simulation of an epidemic spreading on a two-dimensional 100 lattice, where we start from an initial condition with half (randomly chosen) of the individuals being infected (here indicated as black dots). Three examples of the dynamical patterns are shown in the inset

---

[7]The exact form of this probability is obtained as follows. If a given site is empty, and ρ is the fraction of the infected site, the probability none of them will infect is $(1 - \mu\rho)^q$, and thus the probability that some will is $1 - (1 - \mu\rho)^q \approx 1 - e^{-\mu\rho q}$. For small $\mu\rho$ we have $e^{-\mu\rho q} \approx 1 - \mu\rho q$, and a linear relation is obtained.

plots. Using different infection probabilities μ, we measure the probability of propagation by determining if infected individuals can be found after $T = 500$ steps. By averaging over $M = 100$ runs, and using a probability of recovery $\alpha = 0.10$, we obtain a phase transition diagram displaying a sharp change at a given $\mu_c(\Omega) = 0.15$. The simulation thus shows a phase transition between epidemic die out and propagation, but the predicted (mean field) value is different from the $\mu_c = \alpha$ mean field expectation. The reason for this difference is to be found in the strong limitations imposed by local interactions.[8]

The previous result is generic: if local interactions are taken into account, the resulting dynamical patterns and predicted thresholds depart from the mean field approximation (Hinrichsen 2000; Marro and Dickman 1997). If we depart from the lattice model and move into a random graph organization, we recover the basic results predicted by the mean field model. This can be tested by simulating the previous rules on an Erdös-Rényi graph, where the "neighbors" are now replaced by connected nodes in the network. A comparative plot is displayed in figure 5–7. Here a lattice as well as a random graph with $N = 100$ nodes have been used. For the random web we have used a p value that gives an average degree $\langle k \rangle \approx 4$ that compares with the $q = 4$ neighbors of the lattice model. The two curves clearly illustrate the main difference in terms of the speed of propagation of the epidemics. While the lattice model displays a slow, almost linear growth, the random graph allows for a rapid expansion.

An important message (to be reanalyzed at the end of this chapter) is that geography—or its absence—deeply constrains

---

[8]A formal approach to the spatial spread of epidemics on a given domain is to consider a continuous field $\rho = \rho(x, t)$ that follows a partial differential equation

$$\partial\rho/\partial t = \mu\rho(1 - \rho) - \alpha\rho + D(1 - \rho)\nabla^2\rho,$$

where the diffusion coefficient reads $D = \int \mu(r)r^2 dr$ (Kephart and White 1991).

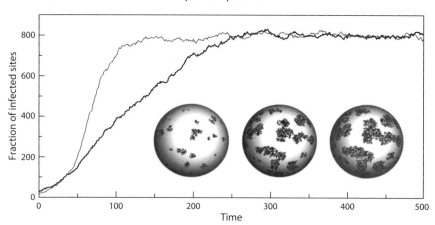

Figure 5.7. The impact of network topology on epidemic spreading. Here two different types of graphs with $N = 100$ nodes have been used. These include an Erdös-Rényi graph where two nodes are connected with a probability p (thin line) to be compared with a square lattice (thick line). Here $q = 4$ and $p = 0.04$, so that the average number of neighbors is $\langle k \rangle = 4$. Three snapshots of the lattice model, with different initial seeds of infected sites, are shown in the inset for three different early stages.

virus propagation. In general, local dynamics leads to propagating fronts (Murray 2004; Shigesada and Kawasaki 1997), and the linear growth displayed by the lattice model was first reported in the growth of viruses on cell monolayers in Petri dishes (Koch 1964). Before transportation was efficient, the role played by geography was also dominant in spreading diseases. Smallpox propagated in waves across the Old World (possibly originating in Egypt)[9] long before it arrived to the New World carried by Cortés and his warriors. In both cases, viruses and other pathogens were carried by humans through old transportation networks. Since then, our society and technology have experienced enormous changes, including the rapid spread of individuals. Geography,

[9]Skin lesions found in mummies are consistent with smallpox symptoms.

in a way, is gone. But we cannot completely ignore it. As climate change modifies our biosphere, the geographic ranges of disease vectors (such as mosquitoes) slowly expands, along with the viruses they carry.

5.4 AIDS: Modeling HIV-1 Transmission

The SIS model is an oversimplified picture of epidemic spreading but it gives us a powerful insight into the presence of eradication thresholds. Moreover, despite its simplicity it is used as a baseline for several well-known epidemics. In chapter 7 we will consider a natural extension of the model incorporating a "recovered" compartment, where individuals who have been infected do not return to the susceptible state. More generally, we might ask how to proceed when the problem under consideration requires more population compartments and thus a larger number of parameters and equations.

In this section we consider a model proposed by Anderson and May (1998) that is aimed at studying the spread of AIDS in a structured population where four compartments are included (figure 5.7). The model assumes a population of $N(t)$ susceptible males that receives an input flow (at a rate β) of new individuals representing a constant immigration. If $X(t)$ stands for the number of susceptibles at time t, the model assumes that infectious males are generated at a given rate λc, where λ is the probability of becoming infected from a partner, i.e.,

$$\lambda = \frac{\eta Y}{N}, \tag{5.13}$$

η being the rate of transmission. This population can now experience two possible transitions: toward either patients developing AIDS or individuals who have antibodies against the virus (seropositive) but show no symptoms of infection and are not infectious. All compartments incorporate a natural death rate μ,

but an additional rate needs to be used for the AIDS patients.
The diagram depicted in figure 5.7 summarizes all the potential
transitions and relevant parameters.

The final model can be described by a set of four differential
equations, namely:

$$\frac{dX}{dt} = \beta - \mu X - \lambda c X \qquad (5.14)$$

$$\frac{dY}{dt} = \lambda c X - (\nu + \mu)Y \qquad (5.15)$$

$$\frac{dA}{dt} = p\nu Y - (d + \mu)A \qquad (5.16)$$

$$\frac{dZ}{dt} = (1 - p)\nu Y - \mu Z; \qquad (5.17)$$

a total population is given by

$$N(t) = X(t) + Y(t) + Z(t) + A(t), \qquad (5.18)$$

which in this case is not constant (since flows include mortality
and thus no strict conservation). In these system, p provides the
rate of seropositives that are infectious, whereas ν will define the
transition rate from infected to AIDS cases. It is important to
realize that λ is given by

$$\lambda = \frac{\eta Y}{X + Y + Z}, \qquad (5.19)$$

and the model also assumes that $A \ll N$.

One specific question related to this model is: can we provide
a reliable estimate of R_0? To answer this question, let us first
determine the sum of all the previous equations, which gives a
simple expression:

$$\frac{dN}{dt} = B - \mu N - dA. \qquad (5.20)$$

Now if we take the previous equation for the evolution of the
number of infectious individuals Y, it is possible to see that, at

time $t = 0$, we can write:

$$\left(\frac{dY}{dt}\right)_{t=0} \approx (\beta c - \nu - \mu)Y. \tag{5.21}$$

This will lead to a propagation of the epidemic if

$$\left(\frac{dY}{dt}\right)_{t=0} > 0 \tag{5.22}$$

and thus if

$$(\beta c - \nu - \mu) = \nu(R_0 - 1) > 0, \tag{5.23}$$

where we have defined

$$R_0 = \frac{\beta c}{\nu} \tag{5.24}$$

as the basic reproductive number. Although other results can also be derived from the model using the appropriate assumptions,[10] we have already obtained an interesting result that, despite the rough assumptions, works very well.

Can we also derive time-dependent estimates of the numbers of individuals developing AIDS? Following the previous approximation, the solution to the approximated equation for dY/dt at early times gives an exponential function,

$$Y(t) = Y(0)e^{\nu(R_0-1)t}, \tag{5.25}$$

predicting exponential growth of the number of infected individuals. If we take this result as a good approximation for the beginning of the outbreak, we can now use the equation for the number of AIDS patients and include the previous result:

$$\frac{dA}{dt} = p\nu Y(t) - (d + \mu)A; \tag{5.26}$$

---

[10]For example, it is possible to provide estimates of the duration of epidemic outbreaks and other dynamical and stability properties.

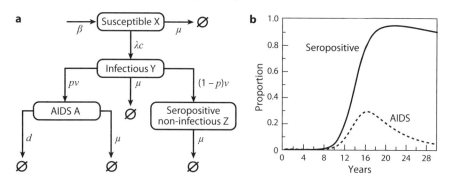

Figure 5.8. Flow diagram of the Anderson and May (1998) model for the development of the AIDS pandemic. Four compartments are included in the model description, as discussed in the text. In (b): Numerical solution of the Anderson-May model for AIDS propagation. Here both seropositive and AIDS populations are displayed (redrawn after Anderson et al. (1986)). The system starts with an initial condition $S(0) + Y(0) = N(0) = 100.000$, and a reproductive rate $R_0 \approx \beta c/v \sim 5.15$ is used. Adapted from Murray (1999).

and if $A(0) = 0$ it is possible to show that the predicted number for $A(t)$ follows

$$A(t) = pvY(0) \left\{ \frac{e^{v(R_0-1)t} - e^{(d+\mu)t}}{r + d + \mu} \right\}. \qquad (5.27)$$

The estimated parameters obtained from existing populations within these groups allowed us to estimate how long it would take to double the number of seropositives, which gave an $\sim$ *nine*-month window. Similarly, the basic reproductive number was calculated to be $R_0 \approx 3 - 4$. An illustrative example of the model predictions is shown in figure 5.8, where the time evolution of both seropositive patients and those developing AIDS is displayed.

The modeling of the AIDS pandemic has been a very active field, and models have been improved to incorporate more detailed information, from additional classes of individuals to age structure (Chavez 1989, 2013; Huang et al. 1992). However, some crucial aspects of the epidemic dynamics require considering an element that has been neglected in the previous models: viruses spread on networks made by individuals who can be connected in complex ways.

5.5 Halting Viruses in Scale-Free Networks

In previous sections we have considered models of epidemic spreading that occur either in a well-mixed (mean field) context, random graphs, or in a spatial domain where interactions are limited to contacts among nearest neighbors on a lattice. The first type deals with a geography-free world whereas the second explicitly takes into account the local nature of the infection process. As discussed in previous sections, the two approaches offer interesting insights, but neither is completely realistic: humans are neither well mixed nor so limited by two-dimensional spatial constraints. In order to fully understand the patterns of epidemic spreading, it is essential to take into account the social networks that underlie human behavior.

What is thus missing from our previous examples? In section 5.4 we have discussed a simple model for the AIDS pandemic, where the population was split into several classes, finding an estimate of R_0. What would be the relevance of adding the structure of human interactions? A crucial finding was to show that sexually transmitted diseases (STDs) seem to propagate across highly heterogeneous networks (figure 5.9a) involving a vast number of individuals having a small number of contacts while a handful have a large number of contacts, thus acting as hubs in the network (Liljeros et al. 2001; Lloyd and May 2002; Schneeberger et al. 2004; Latora et al. 2006).

The resulting degree distribution for these sexual contact networks (figure 5.9b) is a *scale-free* one, and the impact of their heavily asymmetric probability distribution, with a long tail of low-contact cases, deeply changes our perception of disease

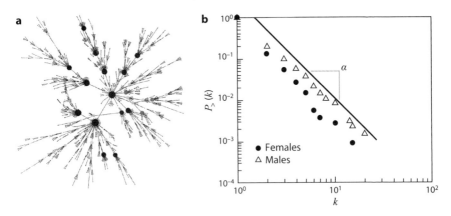

Figure 5.9. Sexual contact networks. In (a) we display an idealized picture of these webs, where most individuals have one or two contacts whereas a handful have many (graph generated using the Barabási and Albert (1999) preferential attachment model; see text). In (b) an example of the observed degree distribution $P_>(k)$ of sexual partners in a given community is shown. Adapted from Liljeros et al. (2001).

propagation and the role played by social interactions. Specifically, scale-free networks (SFNs) display a degree distribution that decays as a power law, namely:

$$P(k) = \frac{1}{Z} k^{-\gamma}, \tag{5.28}$$

with $2 < \gamma < 3$ and Z being a normalization constant. Before proceeding with the analysis of the dynamical process that takes place on a highly heterogeneous scale-free network (SFN), let us consider one possible way of modeling such a network.

The simplest mechanism that allows to build an SFN is the so-called *preferential attachment* rule (Barabási and Albert 1999). In order to build this type of net, we start from a given small network with m_0 nodes, and then apply two simple rules: (i) at each time step, add one new node to the system and (ii) link the new node to m randomly chosen nodes. When choosing the nodes, we assume that the probability $\pi(k_i)$ that a new node will be connected to node i will be proportional to its degree, i.e.,

$$\pi(k_i) = \frac{k_i}{\sum_j k_j}. \tag{5.29}$$

After t time steps a network with $N = t + m_0$ nodes (and mt edges) is generated, exhibiting an SF topology with $\gamma = 3$ (figure 5.8a). Other possibilities include considering aging and cost (Amaral et al. 2000), and can be easily introduced in order to include cut-offs at given connectivities.

The problem of epidemic spreading in SFNs will be explored by means of an SIS model. As before, each node in the graph of interactions will be an individual and each link a potential pathway of disease spreading. The average density of infected individuals, $\rho(t)$ (prevalence), at the mean-field level is

$$\frac{d\rho(t)}{dt} = \mu \langle k \rangle \rho(t) [1 - \rho(t)] - \alpha\rho(t); \tag{5.30}$$

by defining an effective spreading rate $\lambda = \mu/\alpha$, we can simply write:

$$\frac{d\rho(t)}{dt} = \lambda \langle k \rangle \rho(t) [1 - \rho(t)] - \rho(t). \tag{5.31}$$

The benefit of this mean-field equation stems from the fact that density correlations are ignored. On random graphs and related graphs, one can assume that $k \simeq \langle k \rangle$. Following our previous analysis of the contact processes, it is easy to see that a nonzero epidemic threshold exists at $\lambda_c = \langle k \rangle^{-1}$ such that

$$\rho = 0 \qquad \text{if } \lambda < \lambda_c, \tag{5.32}$$

$$\rho \sim 1 - \frac{1}{\langle k \rangle} \qquad \text{if } \lambda \geq \lambda_c. \tag{5.33}$$

So far, everything seems pretty much the same. But a crucial property of SFN changes everything (Pastor-Satorras and Vespignani 2001a, 2001b; Lloyd and May 2001; Dezsö and Barabási 2001; Barabási 2016).

As discussed in section 5.3, simple random graphs have a characteristic average number of links per site that is properly represented by the average degree $\langle k \rangle$. Statistically, that means that the networks are homogeneous and thus they have a well-defined standard deviation $\sigma = \sqrt{\langle k^2 \rangle - \langle k \rangle^2}$ around the mean value. In stark contrast, the fluctuations $\langle k^2 \rangle$ in SFNs diverge for any value $2 < \gamma < 3$, and thus highly connected nodes are statistically significant: the mean field approximation breaks down. In order to take into account these fluctuations, the relative density $\rho_k(t)$ of infected nodes with given connectivity k must be taken into account. The mean-field equations can thus be written as

$$\frac{d\rho_k(t)}{dt} = \lambda k \left[1 - \rho_k(t) \right] \Theta(\rho(t)) - \rho_k(t) \qquad (5.34)$$

for $k = 1, ..., N$. A new term $\Theta(\rho(t))$ indicates the probability that any given link points to an infected node. The probability that a link points to a node with k links is proportional to $k P(k)$. A randomly chosen link is thus more likely to be connected to an infected node with high connectivity, yielding

$$\Theta(\rho(t)) = \frac{\sum_k k P(k) \rho_k(t)}{\sum_k k P(k)}, \qquad (5.35)$$

where $\sum_k k P(k) = \langle k \rangle$ by definition. At the steady state $d\rho_k(t)/dt = 0$, and we get

$$\rho_k = \frac{\lambda k \Theta}{1 + \lambda k \Theta}, \qquad (5.36)$$

and the following relation follows:

$$\Theta = \frac{1}{\langle k \rangle} \sum_k k P(k) \frac{\lambda k \Theta}{1 + \lambda k \Theta}, \qquad (5.37)$$

where Θ is now a function of λ alone.

The solution $\Theta = 0$ is always satisfying the previous equation. A nonzero stationary prevalence ($\rho_k \neq 0$) is obtained when the right-hand side and the left-hand side of equation 5.40, expressed as function of Θ, cross in the interval $0 < \Theta \leq 1$, allowing a nontrivial solution. It is easy to realize that this corresponds to the inequality

$$\frac{d}{d\Theta}\left(\frac{1}{\langle k\rangle}\sum_k kP(k)\frac{\lambda k\Theta}{1+\lambda k\Theta}\right)_{\Theta=0} \geq 1 \qquad (5.38)$$

being satisfied, defining the critical epidemic threshold by:

$$\frac{\sum_k kP(k)\lambda_c k}{\langle k\rangle} = \frac{\langle k^2\rangle}{\langle k\rangle}\lambda_c = 1; \qquad (5.39)$$

in other words, we obtain:

$$\lambda_c = \frac{\langle k\rangle}{\langle k^2\rangle}, \qquad (5.40)$$

which is nothing but the inverse of the coefficient of variation (CV; see Anderson and May 1991). This result means that *in SFNs with $\gamma \in (2,3)$ we have $\lambda_c = 0$, and thus for any λ the infection can pervade the system with a finite prevalence* (Pastor-Satorras and Vespignani 2001). For small λ it is possible to solve explicitly the previous equation and calculate the prevalence in the endemic state as follows:

$$\rho = \sum_k P(k)\rho_k. \qquad (5.41)$$

In the particular case of the Barabási-Albert network with $\gamma = 3$, we find $\rho \sim \exp(-C/\lambda)$ where C is a constant.

The absence of any epidemic threshold in this network can be understood by noticing that in heterogeneous systems the basic reproductive number R_0 contains a correction term linearly dependent on the standard deviation of the connectivity distribution. In SFNs the divergence of the connectivity fluctuations leads to an R_0 that always exceeds unity at any rate λ. This ensures that epidemics always have a finite probability of surviving indefinitely.

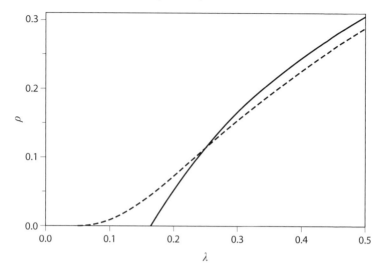

Figure 5.10. Lack of eradication thresholds in epidemic models on scale-free networks. Here the mean field model prediction displaying a phase transition (in an SIS model) is compared with the one shown for a scale-free network (dashed line).

Is there a way out of this bad news? The good news is that a proper assessment of how to prevent hubs from spreading the disease can bring the dynamics back to the threshold-like behavior. This method has been dubbed "curing the hubs" (Dezsö and Barabási 2002), where a cost-effective approach is taken with a policy biased toward hubs. Its authors have proposed a strategy where they assume some degree of accuracy in identifying hubs, measured by means of a parameter ϵ, with curing of a given node v_i being proportional to k_i^ϵ. This factor now will replace the recovery rate α used in our previous models. When $\epsilon = 0$, the curing is random and we recover the previous scenarios, while $\epsilon > 0$ indicates a degree of success scaling with ϵ. It is possible to show that even small values of ϵ allow us to recover the epidemic threshold condition. This can be achieved even if no complete information is available concerning the exact identity of hubs, it is possible to drive the virus into extinction.

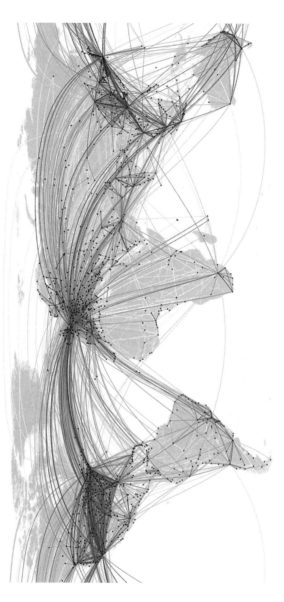

Figure 5.11. Here is a map of around 4,000 airports in the world, with about 440,000 connections between them (picture after Dirk Brokman). As with the sexual contact network, the network is very heterogeneous. Since epidemic spreading can expand into a pandemic event thanks to these webs, identifying the potential role played by hubs and other nodes might be of crucial importance.

The architectures of social and transportation networks, both displaying heterogeneous structures (figure 5.10), represent a threatening risk for the large-scale spread of pandemics, and in this context forecasting and control is a great challenge (Hufnagel et al. 2004; Vespignani 2011). The impact of one of these events can be harmful, even devastating for our society and economy. More importantly, we are pushing the boundaries of our planet to its limits due to a combination of demographic explosion and the overexploitation of and damage to the biosphere. These impacts pervade a crucial aspect of epidemics: how they originate and how they can become pandemic.

6

EMERGENT VIRUSES

As we have discussed in previous chapters, viruses have been major players through evolution, including the origin of some key transitions of life, the history of humankind, and even that of the entire planet, influencing ecological dynamics, geochemical cycles, and biomass turnover. They are not only everywhere, they are also a crucial component of ecological webs. Ecosystems have been deeply modified by humans and, under the right circumstances, these modifications can become a serious threat. Anthropogenic effects and an explosive demography are the triggers of multiple sources of perturbations: habitat loss and fragmentation, species invasions, hunting to extenuation, and extensive agronomical practices that have dramatically changed the landscape are the main examples of such changes. They are occurring at a very fast pace, much faster in fact than any previous biotic change that has taken place on the planet, with the exception of some of the most devastating mass extinction events (LeGuenno 1995; Holmes 2009).

A rather undesirable outcome of such human-driven changes as bringing in close contact farm animals and crops with wild animals and plants that may be reservoirs of a number of viruses is the "emergence" of new pathogenic viruses (Woolhouse et al.

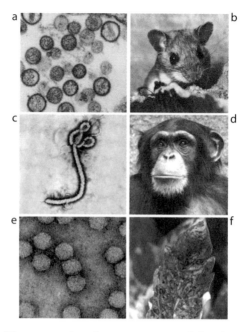

Figure 6.1. Three examples of emergent viruses (left column) and some of their carriers. Hantavirus (a) is one of the recent, most deadly emerging viruses resulting from the jump from its natural carriers (mice; b) to humans. The EBOV (c) may be transmitted to humans through the ingestion of meat from chimpanzees (d). Similarly, the *Tomato torrado virus* (e) is an aphid-transmitted virus that infects tomato plants (f) and was first detected in Spain in 2006.

2001, 2005). Indeed, a virus can be defined as "emerging" if it meets one or more of the following conditions:

1. An already known viral disease that spreads out in a new geographic area or population. Examples of this type of emerging viruses are Zika, Dengue, Chikungunya, West Nile, yellow fever, or the tomato leaf curly geminivirus.
2. A new infection resulting from the evolution of an existing virus that has acquired new biological

properties (e.g., host range, pathogenicity, vector transmission ...). Examples of this second type of emerging virus are many: EBOV (jumping from bats to humans), HIV-1 (from monkeys to humans), the four-corners hantavirus (from rodents to humans), and the canine parovirus (from cats to dogs), and the *Canine distemper virus* infecting sea lions in the North Sea or lions in West Africa; but it also incorporates the periodic rise of new genomic rearrangements of the IAV (H5N1 in Hong-Kong in 1997, H1N1 in Mexico in 2013).

3. A disease or virus that has been infecting us for a long time but remained undetected until diagnosed for the first time due to increased surveillance or to new diagnostic tools. In this sense, all viruses have been "emergent" at some point. Examples of this third class are the human papillomaviruses, the Epstein-Barr (infectious mononucleosis), and the HBV.

The rise of such emergent viruses requires three phases (Woolhouse et al. 2001, 2005). The first phase represents ecological opportunities (section 6.1 below), that is, the chances of the virus spilling over from its reservoir host to a new one. Spillovers can occur with or without changes in the genetic structure or biological properties of the virus, simply as a consequence of increased opportunities of transmission due to perturbations in the ecosystem. However, in many other instances, these early steps in the process of emergence strongly depend on the population genetics of the virus in its reservoir (e.g., how much genetic variability is generated and maintained within the population of the reservoir host), as the likelihood of succeeding in the novel host would depend on the fitness of the different genetic variants maintained in the reservoir host. The second phase is considered to start once the virus infects a novel host and corresponds to all

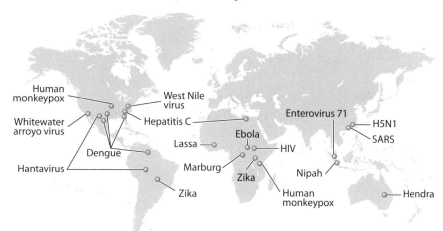

Figure 6.2. Geographic distribution of original outbreaks of emergent (and re-emerging) viruses. Different locations and the associated virus are indicated.

the genetic and evolutionary processes that drive its adaptation to the novel host (section 6.2 below), including changes both in the virus' genome and in its fitness, infectivity, and virulence. This phase usually involves profound changes in the way the virus and the host cells interact. The third and final phase of the emergence process describes the epidemiological spread of the new virus, already more or less adapted, into the global population of the novel host (section 6.3 below). This epidemiological phase is dominated by the demography and behavior (population size and density) and migration patterns, among other factors. Obviously, these three phases are not entirely sequential but can overlap in time and affect each other.

6.1 Ecological Disturbance: Hanta- and Arenaviruses as Case Studies

Many examples of emergent viruses have been reported (figure 6.2). A good example of emergent diseases linked to the breakdown of the equilibrium between a species and its habitat are those caused by the Hanta- or Arenaviruses. Hanta- and Arenaviruses are zoonotic RNA viruses that are transmitted from rodents (e.g., mice and rats) to humans. These rodents are merely a step in the transmission chain between the virus and humans. The relationship between the rodent and the virus is a typical case of commensalism,[1] with the rodents acting as the reservoir host and being unaffected by the virus.

Although direct transmission of the virus (i.e., through bites) is uncommon, infection can occur as a result of eating contaminated food, or touching the mouth or nose after handling a contaminated surface, or even by airborne transmission of the aerosol particles released by the rodents with their saliva. Infection with Hanta- or Arenaviruses can cause quite serious diseases: Hantavirus infection can cause Haemorrhagic Fever with Renal Syndrome (HFRS) or Pulmonary Syndrome (HPS), the latter being a very serious and actually fatal condition in about half the reported cases. Scientists have demonstrated that habitat destruction and climate change can increase the possibility of Hanta- or Arenavirus infection in human populations. The question is: how and through what mechanisms?

It is well known that rodents can thrive in a wide range of habitats, from forest to grassland, canyon, and desert. They can survive in any dry land habitat and invade and exploit disturbed areas. The deer mouse, in particular, is one of the most common transmitters of Hantaviruses to humans; it is extremely common in areas affected by flooding, fire, avalanches, and mining or construction work. Rodents are omnivores and store food for winter consumption, making human-occupied areas especially attractive given the supply of acorns, nuts, insects, other small invertebrates, and various plant parts that rodents need to survive.

---

[1] A type of relation involving two different kinds of organisms, which benefits one and does not harm the other.

Therefore, habitat destruction in the vicinity of suburban areas furnishes an attractive environment to these animals, triggering sizable population outbreaks. These outbreaks are regulated in part by climate change, since variations in temperature and/or rainfall patterns influence rodent populations given the indirect impact on both the total nutritional biomass available (plants, fruits, and invertebrates) and animal reproductive processes (e.g., reproductive seasons or gravidity rates). Hanta- and Arenaviruses are clearly linked directly to ecosystem degradation: changes in habitat or climate have a direct impact on rodent populations and, therefore, eventually cause the emergence of viruses. Nature, however, is not always so evident in terms of cause and effect. As mentioned above, we can expect an extinction cascade to result in the collapse of an ecosystem. As a collateral effect we are also likely to see new species playing an important role in the altered ecosystem, as well as the interaction of new species (and/or habits), thereby opening the door to significant expansion of infectious agents.

6.2 The Genetics of Adaptation to Novel Host

Viruses live in an always fluctuating world (figure 6.3). They move from host to host, sometimes being air- or water-borne but sometimes using vectors (e.g., insects) in which they may or may not reproduce. Within an individual host, viruses face multiple tissues and cell types that differ in physiological and biochemical properties and are constantly being challenged by a diversity of antiviral responses. Some viruses have evolved to be specialized in infecting one or very few host species while others are generalists that successfully infect different host species, sometimes not even related taxonomically. Examples of specialists are Coxsackieviruses, the Epstein-Barr virus, HCV, and the measles and mumps viruses, whose only known mammalian host are humans. Examples of generalist viruses are *Cucumber mosaic*

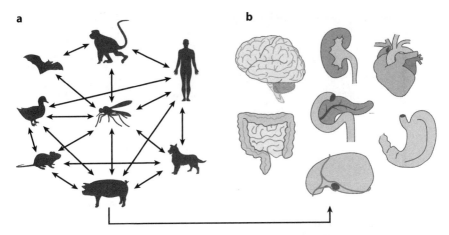

Figure 6.3. Viruses face a world always changing. The figure illustrates the changing environments in which many viruses live. Viruses may move between different hosts, in some instances using vectors into which they may even replicate (as is the case for arboviruses, which use insects as vectors, and the Hanta and Arena-viruses, which use rodents as vectors). Within an infected host, in some instances viruses may show strong tropisms and only replicate in a limited number of tissues and cell types, whereas in other cases, viruses can replicate in different tissues and cell types. Although the image only shows animals, a similar scheme can be drawn for plant viruses.

virus, which infects more than 1,000 species including monocots and dicots, herbaceous plants, shrubs, and trees, and IAV, which infects birds and several different species of mammals. Likewise, some viruses show a strict tissue tropism, infecting only a few cell types, for example, HCV or *Herpes simplex virus*, while others are very generalist and infect multiple tissues, e.g., IAV or EBOV.

By specializing in a single host (species or cells), viruses may reduce competition at the cost of accessing a more limited set of available resources (Futuyma and Moreno 1988). In contrast, a generalist virus would be capable of exploiting multiple hosts, thus enhancing its fitness. Since generalist viruses are not the norm, it is generally assumed that generalism comes with a cost, in keeping with the adage that *a Jack of all trades is a master of none* (Whitlock 1996). Classically, evolutionary biologists justify the evolution of specialization in the existence of fitness trade-offs that limit the fitness of generalists in any alternative hosts (Whitlock 1996; Fry 1996) (figure 6.4a), that is, a generalist will always be a worse competitor on each possible host than the corresponding specialist. What are the causes of such trade-offs? The most obvious mechanism is call antagonistic pleiotropy.

Antagonistic pleiotropy means that mutations that are beneficial in one host may be deleterious in an alternative one (Fry 1996): a mutation in a viral protein that allows interacting with a given host protein may enhance fitness in the actual host, but if the amino acid sequence and structure of the host protein differ somehow among hosts, then the interaction may not work well in alternative ones, and vice versa. A second mechanism that promotes trade-offs is mutation accumulation, in which neutral mutations accumulate by drift in genes that are useless in the actual host but may be essential in a future new one (Kawecki 1994). Both mechanisms involve differences in fitness across hosts; however, they are not equivalent phenomena: natural selection is the only reason for antagonistic pleiotropy, as genetic drift is for mutation accumulation.

6.2.1 Becoming Specialists

A number of evolution experiments with bacteriophages and arboviruses have shown that whenever a virus is adapted to a single novel host type (figure 6.4b), it becomes a specialist that pays a fitness host in alternative hosts. VSV (Holland et al. 1991; Novella et al. 1999; Remold et al. 2008; Turner and Elena 2000; Presloid et al. 200)), *Eastern equine encephalitis virus* (EEEV) (Weaver et al. 1999; Cooper et al. 2001), *Sindbis virus* (SINV) (Greene et al. 2005), and *Venezuelan equine encephalitis virus* (VEEV) (Coffey et al. 2008) have been evolved in and adapted to different lineages of animal cells in *in vitro* cultures.

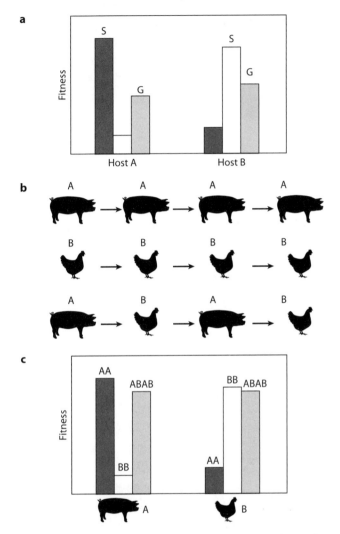

Figure 6.4. Fitness trade-offs across hosts. (a) Expected fitness for specialist and generalist viruses if a trade-off exists. Although both specialist genotypes perform well in their respective hosts, each one is poorly adapted in the other host. In (b) the three classes of experiments are indicated and in (c) is the outcome of three evolution experiments. Viruses evolved in a single host (AAA or BBB) become specialists on their respective hosts; by contrast, a virus evolved in a fluctuating host landscape ($ABAB$) becomes a generalist and improves fitness in both hosts at the same time.

A common result of all these studies is that viral populations evolved on a single cell host type increased fitness in the new host and paid the cost in any alternative host cell type, including the ancestral one (figures 6.4b-c). Likewise, there are several studies with plant viruses in which the same phenomenon has been described. For example, Agudelo-Romero et al. (2008) evolved independent lineages of TEV by serial passages in two different hosts. While TEV lineages maintained in the original tobacco host showed no increase in either viral load or virulence, lineages evolved in the new host pepper showed increases in both traits. However, these increases were specific to the pepper host, and the pepper-adapted lineages did not show any replicative fitness increment in the ancestral tobacco host.

6.2.2 Becoming Generalists

In a second set of studies, the effect of temporal host hetero-geneity has been addressed experimentally (figure 6.4b). Despite methodological differences and the use of different host types, most of these studies came to a common observation: when the viral population alternated in time between two host types, natural selection improved fitness in each type to a similar extent as when adaptation happened to each one individually. For example, VSV populations became generalists without paying fitness costs in any of the alternative hosts (Turner and Elena 2000). However, a significant cost was paid by these general-ist viruses in the ancestral host cell type not included in the fluctuation treatment. The same observation was made for EEEV populations evolved in two alternating cell types (hamster and mosquito): EEEV reached replicative fitness values on each cell type similar to those reached by viral lineages evolved only on single cell types (Weaver et al. 1999).

Therefore, all these results suggest that no fitness trade-off exists when the host landscape fluctuates fast, since the replicative fitness in both environmental extremes is maximized (figure 6.4), potentially generating generalist viruses. However, the observation of a lack of fitness trade-off seems not to be ubiquitous. For example, for some but not all SINV lineages alternatively passaged in mosquito and hamster cells, the replicative fitnesses on each alternative host were lower than those reached by SINV lineages evolved on each host held constant (Greene et al. 2005). This result is still compatible with the existence of a fitness trade-off across hosts. The fact that not all SINV generalist lineages showed the trade-off might be explained by some lineages overcoming the trade-off by finding the right combination of mutations whereas the lineages still showing the trade-off did not find such combinations.

The above discussion concerns evolution in temporally fluctuating hosts. Another highly relevant, and related, issue is evolution in spatially fluctuating hosts. This is, for example, an essential component of mammalian viruses that may use the bloodstream as a highway to spread across different tissues. Yet the question is then, how does spatial heterogeneity affect viral evolutionary dynamics? Cuevas et al. (2006) addressed this question experimentally using VSV. They found that the extent to which VSV adapts to diverse host cell types strongly depends on the migration rate among cell types: increasing migration rate selects for generalist viruses (figure 6.4). By contrast, in the absence of migration, VSV lineages specialized in their local host cell type (figure 6.5).

This result supports the view that migration among hosts must be sufficiently low relative to the strength of selection to generate specialists. Indeed, the conditions for the coexistence of specialist viruses in a heterogeneous host environment are very restrictive. If the selective differences among hosts are not so large, then the balance of production from each host must be roughly equal in order to maintain diversity. This implies that there must be lots of opportunities for generalists to evolve in heterogeneous environments, even if selection favors in the

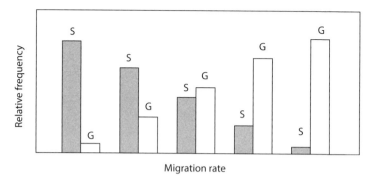

Figure 6.5. Effect of migration rate among different hosts in the fraction of generalist viruses that can be maintained in the population. Migration rate increases from left to right. Increasing migration rate among different cell types favors generalist genotypes (G), whereas a reduced migration rate would favor viral genotypes specialized on each host (S). In the absence of migration, specialist viruses (dark bars) dominate the population. Frequency of generalist viruses (white bars) increases with migration rate.

short term specialization to the host wherein virus productivity is maximized.

Despite their conceptual interest, these studies suffer from the limitation of being undertaken in a highly artificial cell culture system. This limitation has been overcome in recent years by running evolution experiments in whole hosts, which represents a more biologically realistic situation. For example, Coffey et al. (2008) evolved independent lineages of VEEV either in *Aedes aegypti* mosquitoes or rodents, or alternating between both animals. As expected by the trade-off hypothesis, serial *in vivo* mosquito passages resulted in enhancement of mosquito infectivity but at the cost of reduced replication ability in rodents. Consistently, VEEV populations serially passaged in rodents showed an increased replication rate in the vertebrate host but reduced infectivity in mosquitoes. Interestingly, alternating *in vivo* passages between mosquitos and rodents did not increase VEEV fitness in either host.

More recently, evolution experiments done with TEV, in which pepper and tobacco hosts were alternated through time, showed a similar pattern of evolution: viruses evolved in both alternating hosts were as fit on each host as the corresponding specialists (Bedhomme et al. 2012). When only one host was present during the entire evolution period, specialists evolved (see above). Therefore, the conclusions drawn from experiments in cell cultures hold for experiments using whole hosts.

6.2.3 The Causes of Specialization

Above, we have presented two non-mutually exclusive explanations to justify the existence of fitness trade-offs among alternative hosts: antagonistic pleiotropy and mutation accumulation. Given the compactness of virus genomes, with many cases of overlapping reading frames and multifunctional proteins, the former is expected to be a more plausible explanation (Belshaw et al. 2007). Indeed, the experimental results reviewed in the following paragraphs clearly support this as the underlying mechanism for fitness trade-offs. A considerably long list of examples supports the antagonistic pleiotropy mechanism at the molecular level.

A good example of antagonistic pleiotropy was reported in the work with ϕX174 phages adapting to either *Salmonella enterica* or *Escherichia coli* hosts (Crill et al. 2000). The genome of *Salmonella*-evolved ϕX174 phages was fully sequenced and the same two or three substitutions in the major capsid gene were recurrently identified in the different lineages. The fact that independent lineages fixed the same mutation provides strong support for the selective advantage conferred by these mutations in the new host. Indeed, when the *Salmonella*-adapted virus was evolved back on *E. coli*, these mutations quickly reverted to the ancestral stage, thus confirming that these mutations had an antagonistic pleiotropic effect in the *E. coli* host.

A second very remarkable example of antagonistic pleiotropy driving virus specialization to a novel host comes from the plant virus *Pelargonium flower break virus* (PFBV) populations adapted to the experimental host *Chenopodium quinoa* (Rico et al. 2006). PFBV isolates maintained for a long time on *C. quinoa* all fixed five specific amino acid substitutions in the coat protein, which were never found in natural isolates from the natural host geranium. When a wild-type isolate from geranium was mechanically inoculated onto *C. quinoa* leaves, the viral population generated right after the first passage had already fixed two of the *C. quinoa*-specific changes, and after only four additional serial passages the entire *C. quinoa*-specific pattern was fixed in all lineages (Rico et al. 2006). The fact that this pattern has never been found in the natural host, not even incompletely, suggests that it may impose a strong burden for viral replicative fitness on the natural host geranium.

6.3 Epidemics of Emergence

In previous sections we have explored the tempo and mode of emergence of new viruses and the role played by several factors. Different examples have been provided and the adaptation strategies analyzed. Most of these examples and adaptation scenarios describe host-virus interactions that can be understood in terms of the trade-offs and opportunities provided by novel hosts. But there is a global picture that we have not yet addressed concerning the global scale of the phenomenon, namely the population dynamics of emergence. Modeling the rise of new pathogens is a difficult task, since both ecological and evolutionary components are at work, linking the life histories of different species as well as novel forms of interactions. However, a number of general theoretical results can be derived by analysing the conditions under which a virus can successfully expand within a population of susceptible individuals. Such results can help us derive some key relationships between the efficiency of the viral strain (as given by R_0) and the size of the outbreak, thus providing a clue to the requirements for a pandemic to emerge.

In order to predict the size of an epidemic outbreak and how it can depend upon the initial size of the infection, an SIR model will be used. The SIR model (Kermack and McKendrick 1927) involves three distinct subsets, namely S = susceptible, I = infected, and R = recovered. It is somewhat related to the SIS model described in the previous chapter, but the additional compartment makes a difference: instead of returning to the susceptible state, infected individuals overcome the infection but cannot return to the initial, infected compartment. In this case, the reactions defining the SIR model are

$$I + S \xrightarrow{r} 2S \qquad (6.1)$$

for the infection event and

$$I \xrightarrow{\alpha} R \qquad (6.2)$$

for the transition for the recovery process. A constant population $S + I + R = N$ will also be assumed.

The associated differential equations are now:

$$\frac{dS}{dt} = -rIS \qquad (6.3)$$

$$\frac{dI}{dt} = rIS - \alpha I \qquad (6.4)$$

$$\frac{dR}{dt} = \alpha I. \qquad (6.5)$$

This model allows us to make some important predictions concerning the final size of an outbreak of an infectious disease and connect two important parameters: the initial size of the event, I_0 (the number of primary cases), and the basic reproductive number, R_0.

In order to obtain these results, a number of (reasonable) assumptions will be needed. The first is that the initial state (when $t = 0$) is given by $S_0 = N - I_0$ (and a reasonable approximation is thus $S_0 \approx N$, while $I_0 > 0$ and $R(0) = 0$). A first result can be obtained by considering the conditions that allow the epidemic event to start propagating. This requires

$$\left(\frac{dI}{dt}\right)_{t=0} = I_0(r S_0 - \alpha) > 0, \qquad (6.6)$$

and this leads to the condition $r S_0/\alpha > 1$ or, since $S_0 \approx N$, to a basic reproductive number

$$R_0 = \frac{r N}{\alpha} > 1, \qquad (6.7)$$

which sets our definition for this parameter.

Our goal here is to know how many people get infected through the outbreak, which means counting the final (steady) number of recovered individuals, to be indicated as R_∞. Using the relation

$$\frac{dS}{dR} = \frac{\left(\frac{dS}{dt}\right)}{\left(\frac{dR}{dt}\right)} \qquad (6.8)$$

it is easy to see that this leads to

$$\frac{dS}{dR} = -\frac{R_0}{N} S. \qquad (6.9)$$

This can be integrated to give an exponential decay

$$S(t) = S_0 \exp\left(-\frac{R_0}{N} R\right). \qquad (6.10)$$

If we include this in the equation for dR/dt we have now:

$$\frac{dR}{dt} = \alpha \left(N - S - R\right) \qquad (6.11)$$

$$= \alpha \left(N - (N - I_0) \exp\left(-\frac{R_0}{N} R\right) - R\right). \qquad (6.12)$$

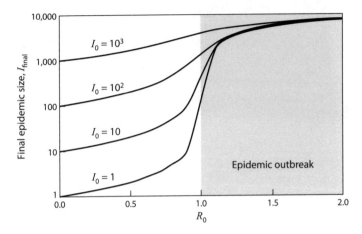

Figure 6.6. Impact of the basic reproductive number R_0 on the population dynamics of pathogen emergence. (a) Relationship between final epidemic size (log scale) and R_0 for a total population $N = 10,000$, for different initial infection size I_0.

The outbreak will end once the condition $dR/dt = 0$ is achieved, which gives our final result for the final value for R_∞:

$$R_\infty = N - (N - I_0) \exp\left(-\frac{R_0}{N} R_\infty\right). \qquad (6.13)$$

The previous equation gives us an implicit relation that measures the impact of the pathogen efficiency (as measured by R_0) on the population dynamics of viral emergence (Woolhouse et al. 2001) and its dependence on I_0. The curves displayed in figure 6.6 illustrate the role played by the initial infection combined with the potential for transmission. For values of R_0 below the critical $R_0 = 1$ we can see that small initial infections barely propagate through the total population, and infections are directly acquired from the primary cases.

Instead, for $R_0 > 1$ this limited spread will be replaced by a major epidemic (gray area), and most infection events are a consequence of propagation from new infected hosts. Examples

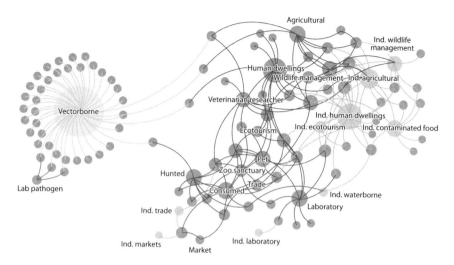

Figure 6.7. A bipartite graph linking emergent viruses transmitted from wildlife to humans and different classes of high-risk interfaces. The latter are displayed as circles with varying size and labeled. Viruses capable of exploiting diverse interfaces appear more connected. After Kreuder Johnson et al. (2015).

of the first category are avian influenza or monkeypox, whereas the second include HIV-1, IAV, and the SARS coronavirus (Woolhouse et al. 2005). As we already discussed in the previous chapter, this is a critical change that has an important implication: any given pathogen slightly below $R_0 = 1$ can become an emergent one by acquiring a small advantage that allows it to cross the critical boundary.

The challenge of predicting, preventing, and fighting emergent diseases, particularly those of viral origin, is enormous. Understanding emergent viruses requires an integration of ecological, epidemiological, social, cultural, and engineering components. Multiple interfaces between humans and pathogens exist that can facilitate the emergence of viral novelties. In figure 6.7 a network has been constructed linking known zoonotic viruses (circles with no labels) and potential high-risk interfaces of disease transmission (Kreuder Johnson et al. 2015). This graph displays high-risk interfaces whose size is proportional to the number of viruses reported for each transmission interface, including direct and indirect contact as well as transmission by vector (large node on the left). Virus node size (here 86 viruses were studied) reflects the number of connections to different transmission interfaces. Larger numbers of connections indicate a higher ecological plasticity of viruses through use of multiple opportunities for transmission.

As discussed in Kreuder Johnson et al. (2015), more dedicated research of the epidemiology of zoonoses at high-risk human-animal interfaces is needed to assess risk of viral disease emergence and direct global, as well as local, disease prevention and control. Ongoing studies indicate that the risk for new virus emergence is higher at those interfaces that facilitated disease threats in the past. Unfortunately, wild animal hosts and high-risk interfaces facilitating spillover of zoonotic viruses, particularly beyond their first emergence, remain vastly underreported. In this area, mathematical and computational models could play a key role.

7

ORIGINS

7.1 Are Viruses Inevitable?

The two most controversial questions that exist in evolutionary virology are about the origin of viruses (Forterre and Prangishvili 2009; Koonin and Dolja 2013) and whether viruses are alive or not (López-García and Moreira 2009; Villarreal 2004). No agreement exists among researchers for either of them. The second question, in our opinion perhaps more semantic than real, is strongly dependent on the first. The lack of agreement between scientists on virus origins is reflected by the number of theories brought forward to explain virus origins. In this chapter we will review the most popular and up-to-date theories without taking a position in support of any of them. By contrast, we do take a strong position on the question of whether viruses are alive or simply large molecular complexes.

The parasitic lifestyle is one of the most represented among life forms. Parasites exist at any level of biological organization: from the molecular to the organismal levels. They pervade all biology (and even beyond that; see chapter 8). Parasites exist infecting from viruses (yes, also viruses!) and prokaryotes to across all the degrees of complexity of eukaryotes. Therefore, parasites are doubtless the most successful life strategy on the planet, taking merciless advantage of every known living creature. Likewise, parasites can be from all levels of biological complexity: viruses, bacteria, archaea, fungi, protozoa, and even plants and animals can become parasites. However, the parasites that have had the biggest impact on evolution are the so-called microparasites: viruses, bacteria, fungi, and protozoa can shape the evolution of their hosts by allowing certain genotypes to survive infection while others have a very reduced fitness.

A good illustration of this selection of host genotypes comes, for example for the case of humans, from heterozygous alleles for the sickle cell anaemia being more resistant to infection by the malaria parasite than individuals homozygous for the wild-type allele (Motulsky 1964). This selection is still happening today, despite the tremendous advances in antimicrobial therapies witnessed in the last century. HIV-1 and tuberculosis, for instance, are driving evolutionary change in parts of our genome, such as the immune system genes (Kloverpris et al. 2016; Perrin 2015). Obviously, the evolutionary relationship between hosts and parasites is bidirectional: hosts can influence the evolution of their parasites too. For example, diseases that require direct contact for transmission often evolve to be less deadly, as for example the well-known case of myxomavirus infecting rabbits (Kerr et al. 2015), ensuring a host will at least live long enough to pass it on. Viruses such as influenza are strongly engaged in an arms race with our immune system (Yoon et al. 2014; Hannoun 2013). New escape mutants emerge every year that escape from our immune systems (and vaccines).

Parasites can also drive the evolution of host genomes at a more basic level. For example, parasitic fragments of DNA, called transposons, which can cut and move themselves all over the host genome, can be transformed into new genes, alter the expression of other genes wherein they land, or favor the fixation of new mutations and chromosomal rearrangements that add to the genetic variation of host populations (Krupovic and Koonin

2016). Lastly, but not less interestingly, parasites have even been implicated in the origins of sex, as they may have been driving selection of mechanisms to generate new genetic variation that may help hosts escape from recognition by their parasites (Lively 2010).

7.2 Evidence from Digital Evolution

In previous chapters we have studied a number of theoretical and computational approaches to the modeling of virus dynamics and how viruses coevolve with their potential hosts. In all these models, both viruses (parasites) and their hosts are already present from the beginning. However, we should consider these models under a more general framework in which parasites can emerge. One of the relevant questions that can be formulated here is whether virus-like agents can emerge in other contexts different from biology. One example that we will explore in chapter 8 is computer viruses. They can be seen as a negative effect of the rise and expansion of computers since the personal computer revolution that started at the end of the 1970s. Although—as discussed in chapter 2—the IT revolution started in the 1950s, a crucial component of the information infrastructures that had to be in place for the emergence of computer bugs was the presence of the right information-storing systems (Augarten 1984; Gleick 2011; Dyson 2012) and, more importantly, a network that connects computers. As also happens in an epidemic spreading scenario (see chapter 5), infection will not take place unless individuals interact, making effective spreading of the pathogen possible.[1]

---

[1]In this context, it has been claimed that the telegraph was the *Victorian Internet*, which shared many surprising commonalities with the modern world wide web. However, a major difference between the two was the lack of information storage in the old version, preventing the Victorian web from the development of viruses but also from becoming a true information network.

Some insight into the origins of viruses can be obtained, once again, from the analysis of evolving artificial life systems. Some early attempts at creating digital organisms were undertaken by Niles Barricelli, a Norwegian-Italian researcher formerly trained in genetics of viruses and with some background in physics, who was invited by John von Neumann to work at the Institute for Advanced Studies. ENIAC, the largest computer at the time, provided an unexpected arena for making the first steps toward artificial life (Barricelli 1962, 1963). In a surprisingly visionary way, Barricelli aimed at exploring evolution by supplying a virtual world of bits where the first digital creatures emerged from a simulation. Barricelli's programs were confined to the low-memory domains of ENIAC, but even with such limitations he was able to prove how complex interactions and innovations could result from a virtual evolution experiment. The simulations involved a one-dimensional world not too different from the cellular automata models invented a few years before (in 1940) by Stanislaw Ulam and von Neumann. The universe was limited to a line of 512 "genes" that could adopt a range of integer values. By defining a simple set of rules for mutation and reproduction, new numbers would be generated, and the process was repeated again and again. Among Barricelli's findings was the emergence of simple codes acting as parasitic entities that sometimes spread exponentially and eventually filled all the available memory space, which was followed by a subsequent extinction of the entire ecosystem.[2]

The idea of evolving virtual creatures in the computer and its potential implications lay dormant for decades until the 1990s, when ecologist Tom Ray found that—among other things—parasitic programs were an inevitable outcome of digital ecologies

---

[2]Actually, it is interesting to notice that, in order to prevent parasites (especially the smaller ones) from spreading and overloading the system, Barricelli made up some additional ad hoc rules to limit the spontaneous emergence and spread of viral-like entities (Dyson, 2012).

(Ray 1992, 1994; Adami 1998). Driven by the question of how diverse ecosystems emerge and persist, he designed a digital model of evolving species based on a set of computer programs competing for the computer memory and displaying mistakes while copying themselves (Ray 1992) to be stored in available memory positions. Under the constraints imposed by finite computer resources, the so-called Tierra model was able to show how some evolutionary innovations can spontaneously develop. In particular, some major transitions took place as soon as programs started to compete.

An early event was a genome reduction innovation, related to the fact that shorter programs can replicate faster than larger ones. This occurs when parts of the coded program can be removed with no consequences. In that respect, redundant pieces of code could be deleted with no harm. Later on, shorter programs emerged, unable to replicate themselves. In other words, parasites came to (digital) life. Hyperparasites, that is, programs able to replicate using pieces of code carried out by parasites, came later, and some programs developed the capacity for exchanging parts of their codes, mainly as a response to escape from parasites (Hamilton et al. 1990; Hillis 1990), thus defining an innovation that we can label as a primitive version of sex (Ray 1991, 1994, 1998). Eventually, groups of slow replicating programs were able to replicate faster through cooperation.

Since these early efforts, many other studies have been devoted to the study of the evolution of artificial host-parasite systems. Most of these models assume the presence of a distinction between the two classes of agents, but not so many have explored the emergence of parasites. One elegant illustration of a model aiming at explaining the emergence of viral entities is provided by the work of Paulien Hogeweg and co-workers (Takeuchi and Hogeweg 2008; Colizzi and Hogeweg 2016).

One especially simple example is where agents evolve cooperation in a two-dimensional lattice (Colizzi and Hogeweg 2016).

The model introduces the use of public goods produced by agents, which on one hand share the public good and use it for reproduction, and on the other compete for existing available sites. Reproduction takes place with mutation and the result of this process is the speciation of a cooperative and a selfish lineage.

Specifically, these authors use an $L \times L$ square lattice $\Omega = \{r = (i, j) | 1 \leq i, j \leq L\}$, where each site $f \in \Omega$ can be occupied by an individual—$S(r) = 1$—or empty—$S(r) = 0$. Each individual produces a good with rate $\phi(r)$ per time step, which is shared with all q neighbors in a neighborhood $\Gamma(r)$. The reproduction probability of each occupied site—provided that an empty site is present within the neighborhood—defines the fitness $f(r)$ of the individual. The benefit $B(r)$ is given by the availability of the public good, i.e.,

$$B(r) = b \left[\frac{\phi(r)}{q + 1} + \frac{1}{q + 1} \sum_{u \in \Gamma(r)} \phi(u) \right], \qquad (7.1)$$

where b is the benefit per unit of the public good and $q = 8$ nearest neighbors were used in the original study. There is a cost $\rho(r)$ for each site producing the public good, which is taken as proportional to production, i.e., $\rho(r) = c\phi(r)$. The fitness $f(r)$ associated to a given (occupied) site will be the difference between cost and benefit, i.e.,

$$f(r) = \Theta(B(r) - C(r)), \qquad (7.2)$$

where $\Theta(z) = z$ for $z > 0$ and zero otherwise. The total fitness of a set of m neighboring individuals competing for a given empty site will be

$$F = f(r) + \sum_k f(k), \qquad (7.3)$$

and a given site with $S(r) = 1$ will occupy the adjacent empty site with a probability (of replication)

$$R(r) = \frac{f(r)}{F}\left(1 - e^{-F}\right). \qquad (7.4)$$

Here the term in parentheses provides the probability that at least one site replicates. This is consistent with the intuition that replication is more likely to occur near sites with higher availability of public goods.

The model is completed by specific rules introducing mutation: each time a replication occurs from a site $r \in \Omega$, the new site $k \in \Gamma(r)$ will have a different $\phi(k)$ value. This happens with a probability μ, and the new value will be $\phi(k) = P(r) + \xi$, with ξ a random number within the interval $[-\delta/2, \delta/2]$. The model is completed with movement and death rules: individuals can randomly move to nearest positions and also die with probabilities k_{mov} and k_d, respectively.

The generic result of this model and its variants is that there is always an evolutionary bifurcation (figure 7.1a) between the two classes of replicators. The upper and lower branches are associated with cooperators producing large amounts of public goods and parasites exploiting cooperators (and thus showing low production levels), respectively. The gray scale provides an estimate of the population size. The spatial dynamics of this system is exemplified in the sequence of snapshots (figure 7.1b-g) where complex spatial structures arise as the two populations coevolve, with cooperators forming coherent waves followed by parasites (dark borders surrounding the waves) that will exploit the public good. These parasites appear aggregated on the edge of cooperating clusters, whose stability is undermined by the cost of cooperation. Despite the cost of maintaining cooperation, this model shows that cooperators and parasites both emerge, with the latter favoring the increase in cooperation of the former.

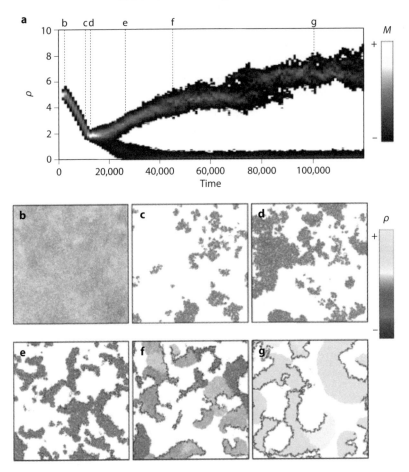

Figure 7.1. Evolution of replicators and viruses in a model of public good production (adapted from Colizzi and Hogeweg (2016)). Here populations evolve on a two-dimensional lattice where they share a public good and compete for available empty sites. Individuals produce a public good at a rate p that spreads evenly over the local neighbors, conferring a benefit on all sites affected. A price is also paid by individuals for the production of the public good. Along with mutation and some stochastic effects, both cooperators and parasites evolve with a characteristic bifurcation (a). In (b-g) several snapshots of the system (indicated above) are displayed, showing how cooperator fronts spread while followed by a front of parasites. Here $b = 10$, $\mu = 0.05$, $\delta = 0.1$, $k_{mov} = 0.02$, and $k_d = 0.2$.

This is a highly simplified model, but provides a powerful illustration of the role played by the interplay between mutation and local dispersal, which is considered a crucial component in the evolution of early cells and viruses (Koonin et al. 2006).

7.3 Where Do Viruses Come From?

Viruses are found infecting all forms of life and have probably been around since the first cells arose, or perhaps even before them. Tracing back the origin of viruses is a titanic, almost impossible, endeavor because they do not form fossils, and the only sources of information are molecular phylogenies and comparative techniques that have been extensively used to compare the DNA or RNA genomes of today's viruses, and reconstruct backwards their evolutionary history, hopefully, to reach their origins. It has been often stated that viruses are polyphyletic, i.e., that different viral lineages originated independently and thus they have no single origin. In particular, RNA and DNA viruses were thought to be evolutionarily unrelated. However, the overall structural similarity between the viral proteins forming the envelopes and capsids suggests at least a common mechanism for their appearance (Abrescia et al. 2012; Krupovic and Bamford 2010). Three main hypotheses have been brought forward to explain the origins of viruses (Forterre 2006a; Forterre and Prangishvili, 2009).

7.3.1 Regressive Hypothesis

This is also called the degeneracy hypothesis, or the reduction hypothesis. The regressive hypothesis suggests that viruses may have once been small cells that parasitized larger cells. As time went on and the parasite became more dependent on the host cell to complete its life cycle, genes not strictly necessary for their acquired parasitism were lost. This hypothesis is grounded on the observation that bacteria such as *Rickettsia*, *Chlamydia*, and

Buchnera are living cells that, like viruses, can reproduce only inside host cells and their dependence on their intracellular life has resulted in the loss of genes that enabled them to survive outside a cell. However, this hypothesis was classically attacked because of two main weaknesses: (i) intermediate forms between cells and viruses were not known and (ii) parasites derived from cells in the three domains of life (e.g., *Mycoplasma* in Bacteria, *Microsporidia* in Eukarya, and *Nanoarchea* in Archaea) have retained their cellular character despite extensive periods of evolution as parasites.

7.3.2 Cellular Origin Hypothesis

This is sometimes called the vagrancy hypothesis, or the escape hypothesis. The second classic hypothesis for the origin of viruses states that some viruses may have evolved from pieces of DNA or RNA that "escaped" from the genome of cells. The escaped DNA could have come from plasmids (pieces of naked DNA that can move between cells) or from transposons (molecules of DNA that replicate and move around the cellular genomes to different positions). Called *jumping genes* when discovered in corn by Barbara McClintock in the 1950s, transposons are a broad class of mobile genetic elements that parasitize the genome of their hosts and encode for very few proteins, mostly for the transposase enzyme responsible for their movement.

This hypothesis became popular at some point partly because of the dissatisfaction with other explanations; partly because it was a priori supported by the observation that present-day viruses can integrate cellular genes into their own genome (e.g., the heat shock protein 70 homologous gene inserted in the genome of plant closteroviruses (Dolja et al. 2006), or the host transposons inserted in the baculovirus genomes (Gilbert et al. 2014)); and partly because a particular class of transposons, the retrotransposons, are structurally very similar to the retroviruses (Hull and Will 1989), although in this particular case the situation can be seen from the opposite perspective and consider the retrotransposons as retroviruses inserted long ago that have lost some of its functions, notably the ability to encapsidate and transmit among cells.

The cellular origin hypothesis has several drawbacks as well. (i) It does not specify how a free nucleic acid could have recruited a capsid and the complex mechanisms required to deliver its content inside a host cell. In this context, it is worth noting that viral capsids do not share any structural or sequence resemblance with cellular components. (ii) The hypothesis predicts that viruses infecting the three domains of cellular life would have originated within each domain, that is, bacteriophages would have originated from bacterial genomes whereas the origin of eukaryotic viruses would be the genome of a eukaryotic cell. Therefore, one may expect to find similarities between viral proteins encoded by viruses from one domain and their cellular homologues in that domain, but homologies between viruses from different domains cannot be expected.

Reality has been perverse with this hypothesis: the similarities between viral genes from different domains are larger than they are among viral genes and their corresponding host cell genes. Of course, in a few cases, viral proteins resemble homologous proteins encoded by their hosts, indicating a recent transfer of these proteins from cells to viruses. From this observation, it has been argued by defendants of the escape hypothesis that all viral proteins must have a cellular origin, and their "ancestrality" is just an artifact generated by phylogenetic methods due to the fast rates of evolution shown by viral genomes. However, obviously, this explanation fails to account for the vast majority of viral proteins without cellular homologues.

7.3.3 Protobiont Hypothesis

This is also called the virus-first hypothesis and suggests that viruses may have evolved from complex molecules of protein

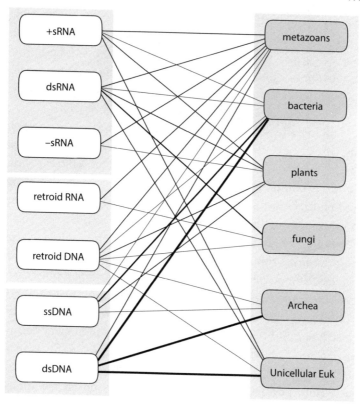

Figure 7.2. Viruses and their host niches. Here we display a bipartite graph including the different classes of viral genomes (left) and their potential host niches (right). The lines indicate known interactions and their thickness provides a relative measure of strength, with thin and thick links corresponding to rare and common virus-host interactions respectively. Based on Koonin et al. (2006).

and ribonucleic acids at the same time as cells first appeared on earth, and would have been dependent on cellular life from the very beginning. In the primitive precellular soup, as in any other replicating system, parasites would had also evolved that grew at the expenses of other more complex molecular systems. When cellular membranes first arose and complex replicative hypercycles isolated themselves from the environment by acquiring membranes, these innovations pushed molecular parasites into strong selection to acquire the ability to cross the primitive membranes. In this context, mathematical and computer models support the idea that complex spatial structures (not necessarily vesicles) can make a big difference (as discussed above).

This hypothesis was dismissed for a long time since all present-day viruses are obligatory parasites requiring an intracellular development stage for their reproduction. This "ancient virus world" hypothesis implies that the primordial origin of diverse viral replication-expression strategies coevolved with the increase in complexity of their corresponding hosts. Positive-sense RNA viruses are thought to be the most ancient type and evolved within the primordial soup, along with the primitive RNA-based cells. The dsRNA and negative-sense ssRNA viruses, which carry their replication machinery within virions, are then most likely to have evolved later on, derived from the positive-sense ssRNA viruses. Indeed, negative-sense ssRNA viruses probably originated during the radiation of primitive arthropods (Li et al. 2015), with a later jump to the plants on which these arthropods feed. Under this vision, positive-sense ssRNA viruses are indeed the direct descendants of the primordial RNA-protein world, whereas the reverse-transcribing viruses provide a possible intermediate for the transition to the DNA world.

Therefore, the central debating point in the discussion about the origin of viruses is whether they are ancient, first appearing before the last universal cellular ancestor (LUCA), or have evolved more recently, such that their ancestry lies with genes that escaped from the genomes of their cellular host organisms. Studies of viral origins have been bolstered in the last two decades by two remarkable observations: the discovery and genome sequencing of giant viruses and the report of apparent homology between the capsid architectures of viruses that possess no primary sequence similarity. As Patrick Forterre (2006b) has pointed out, we cannot

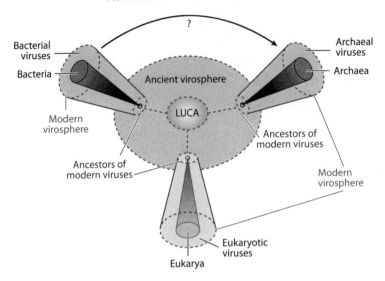

Figure 7.3. Redefining viruses. Here we represent the three domains of life (that have evolved from a LUCA) along with viruses with their capsids. Newly defined viruses have a capsid but no ribosome (modified from Raoult and Forterre (2008)).

pretend to understand the origin of viruses from the perspective of the modern biosphere (i.e., modern viruses infect modern cells; modern cells cannot regress to viral forms; free DNA cannot recruit proteins for encapsidation; etc). Therefore, the three hypotheses must be revisited by considering that viruses originated before LUCA.

The idea of RNA viruses, and even more clearly of viroids,[3] being ancient was easily accepted in the context of the RNA World theory (Gilbert 1986). It has been convincingly argued by several authors than RNA viruses and viroids could be relics of a pre-DNA world in which organisms, even primitive cells, had RNA as the only carrier of genetic information, and proteins as machines

---

[3] Subviral pathogens of plants composed only by a really small noncoding circular molecule of RNA (Flores et al. 2014).

to ensure the transmission of this information. Retroviruses would be relics of the RNA-to-DNA world transition. A priori, the idea that RNA viruses preceded cells in the history of life may sound weird since they are most commonly found infecting eukaryotic cells, although both ssRNA and dsRNA viruses are found infecting bacteria. More interestingly, bacteria-infecting and eukaryote-infecting dsRNA viruses share strong structural similarities and life cycles, thus supporting a common origin before the separation of both cellular domains.

The pre-cellular origin of DNA viruses was also postulated by Forterre (2006a, 2006b) on the basis of the sequence similarity (or lack there) between bacteriophage T4 type II DNA topoisomerase and bacterial girase and eukaryotic type II DNA topoisomerase. Likewise, human Adenovirus and *Bacillus subtillis* bacteriophage $\phi29$ use a similar atypical protein-priming mechanism to replicate their DNA (nonexisting in the cellular world) and encode for a unique type of DNA polymerase that can use such a template to initiate its own replication. These proteins were clearly of neither bacterial nor eukaryotic origin but representative of a new domain that existed before the separation of the three cellular domains.

Viroids are molecules of RNA that are not classified as viruses because they lack a protein coat. However, they have characteristics that are common to several viruses and are often called subviral agents. Other subviral agents are the RNA satellites, hyperparasites that parasitize an RNA virus (Palukaitis 2016). Although RNA satellites are more common among plant RNA viruses than among animal RNA viruses, a particularly well-known case of such satellites is the *Hepatitis delta virus* of humans, which has an RNA genome similar to viroids but has a protein coat derived from its helper virus, HBV, and cannot produce one of its own (Littlejohn et al. 2016). It is, therefore, a defective virus and cannot replicate without the help of HBV. A particularly interesting example of DNA satellites is the virophage that infects giant mimiviruses (Zhou et al. 2013) that are parasites of amoebae; its prototype is the Sputnik virophage that parasitizes the giant *Acanthamoeba polyphaga* mimivirus (La Scola et al. 2008) (already presented in chapter 2). Sputnik has a ca. 18 kb circular dsDNA genome, like a huge plasmid, and, amazingly, it encodes for proteins with homology to the three cellular domains as well as proteins homologous to ATPases from bacteriophages and eukaryotic viruses. Satellites may represent evolutionary intermediates between viroids and viruses or, more likely, other remnants of the RNA world.

The paradigm that viruses have small genomes and are relatively simple in comparison to cellular life has been overturned with the discovery of giant viruses (already presented in chapter 2), larger than some of the smallest bacteria (e.g., *Mycoplasma genitalium*). Over a decade ago, Raoult et al. (2004) characterized the 1.8 Gb genome of the first mimivirus. This genome contained more than 900 putative genes, some resembling nonviral genes involved in translation and protein production. Two contending hypotheses have been brought forward to explain the origin of these genes. Firstly, they could have been acquired from their cellular hosts. Second, mimiviruses may have descended from a free-living cell that gradually lost most of its genes as it became a parasite (according to the cellular origin hypothesis). This mimivirus precursor may represent a new branch of the tree of life, one predating the emergence of the three major branches (bacteria, archaea, and eukaryotes).

Since the discovery of the first mimivirus, the brotherhood of giant viruses has been enlarged with members having larger and larger genomes, the largest one described so far being the giantic pandoraviruses (named so because of their amphora shape and the surprises they contain) (figure 7.4) (Philippe et al. 2013). They have strikingly different genes and physical appearances from other viruses. After being discovered in the late nineteenth century, viruses were quickly demoted to inert particles, too simple to be living beings: no more than a protein

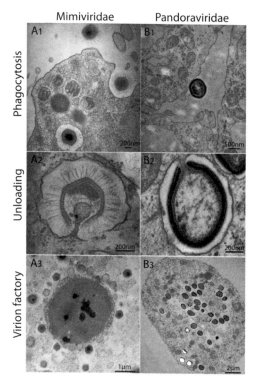

Figure 7.4. Electron microphotographs comparing mimivirus and pandoravirus mechanisms of cell entry, uncoating (and shape), and formation of cellular factories. Both classes of giant viruses largely differ in these steps and structures.

package enclosing a tiny genetic material without any metabolic capability. The giant viruses have destroyed this naive perspective of the viral world. Giant viruses thus can be viewed as an intermediate form between a true cell and a virus. Indeed, the mimivirus genome encodes several proteins that are also present in the three domains of life, and a phylogenetic tree places the mimivirus proteome in between Archaea and Eukarya, as expected for a representative of a fourth domain with no longer existing free-living representatives. In the following sections we will discuss some of the evolutionary implications of these new giant viruses on the evolution of the eukaryotic cell.

As already mentioned, there are problems with all the hypotheses for the origin of viruses: the regressive hypothesis did not explain why even the smallest of cellular parasites do not resemble viruses in any way. The escape hypothesis did not explain the complex capsids and other structures on virus particles. The virus-first hypothesis contravened the definition of viruses in that they require host cells. Nonetheless, viruses are now recognized as ancient and as having origins that predate the divergence of life into the three domains. This discovery has led modern virologists to reconsider and reevaluate these three classical hypotheses. The evidence of an ancestral world of RNA cells and computer analysis of viral and host DNA sequences are giving a better understanding of the evolutionary relationships between different viruses, and may help identify the ancestors of modern viruses. To date, such analyses have not proved which of these hypotheses is correct (Koonin et al. 2006; Koonin and Dolja 2013). However, it seems unlikely that all currently known viruses have a common ancestor, and viruses have probably arisen numerous times in the very remote past by one or more mechanisms.

7.4 Viruses and the Origin of Cells

An obvious question emerging from the RNA world hypothesis is about where DNA comes from. Forterre (2002) proposed that DNA was a viral invention to gain resistance against the RNA-specific defense mechanisms that at that time may have evolved in the RNA host cells. By incorporating this chemical modification into their genetic code, the newly emerged DNA viruses had a clear selective advantage. This is supported by the fact that most modern DNA viruses encode for ribonucleotide reductases and thymidylate synthases enzymatic activities required to produce DNA precursors (Myllykallio et al. 2002). To explain how

DNA replaced RNA chromosomes in the primitive cells, Forterre argues that DNA viruses in persisting infections (non-lytic) lost genes encoding their capsid proteins and lytic functions, thus becoming something similar to a DNA plasmid in an RNA cell. These plasmids could then increase in size by acquiring host RNA genes by the action of a reverse transcriptase enzyme. If the DNA plasmid was replicatively more efficient and stable, it would be to the benefit of the RNA cell to transfer all its genes into this new DNA chromosome, thus slowly replacing RNA chromosomes by DNA ones. The larger stability of DNA molecules compared to RNA molecules opened the door for increasing the size of chromosomes to levels impossible to attain by primitive RNA chromosomes, thus opening the possibility for increasing complexity.

Furthermore, Forterre (2006) argues that given the differences among Bacteria, Archaea, and Eukarya in their replication and translation machinery, it is likely that three independent events of association between large-DNA viruses and RNA-cells took place. These three symbiotic events resulted in the three cellular domains of life that we know now, all three processes taking place in a relatively short period of time or in isolation, so they outcompeted their surrounding RNA cell ancestors. Consequently, additional possible domains have not survived nowadays (or have not been discovered yet . . .). Another explanation for why there are only three cellular domains is that the events described in the previous paragraph are so rare and complex that the probability of more cases rendering additional cellular domains would be too low.

The above theories explain the origin of DNA as the contemporary genetic material of a cell, but do not explain the origin of the eukaryotic nucleus. Bell (2001, 2006) proposed the viral eukaryogenesis theory to explain the origin of the complex eukaryotic nucleus as the result of a consortium between an archaeal cell that provided the cytoplasm, a bacterium that provided the mitochondrion, and a large DNA virus (possibly mimivirus-like) that provided the nucleus. In a process similar to those described in the previous paragraph, the DNA virus genome acquired genes from both the bacteria and the archaea and took over the role of information storage for the consortium. The archaeal host retained its function of gene translation and of general metabolism upon transferring the relevant genes into the viral chromosome; the bacterium retained its ability to anaerobically produce energy (ATP) and transferred most of its functions to the viral chromosome. Bell also proposes that mitosis, meiosis, and sexual cycles arose as a consequence of selective pressures upon the lysogenic virus to maintain itself at a low copy number (not blowing out the consortium) while still being capable of transmitting into the population.

7.5 Viruses as Sources of Evolutionary Novelties

It should be clear now that viruses had an essential role in the early evolution of life and the origin of eukaryotic cells. However, the impact of viruses on evolution not only as harmful parasites goes beyond the origins of cellular domains of life and extends to the acquisition of many evolutionary novelties. For example, in eukaryotes, sequences derived from transposon mobile elements and endogenous retroviruses can account for at least 50% of mammalian genomes and up to 90% of plant genomes (Lynch and Conery 2003). Although integrated transposons are the subject of strong functional suppression by eukaryotes, they are potentially the source of new cellular genes by exaptation[4] (Freschotte and Pritham 2007). Examples of viral genes impacting cellular evolution include bacteriphages that provided their bacterial hosts with powerful toxins that bacteria possibly originally used to fight against predator protists (Wagner

[4]A change in the function of a trait during evolution.

and Waldor 2002). Integration of *Bracoviruses* in wasp genomes allowed the larvae to feed on arthropod hosts using viral-encoded proteins to manipulate their victims (Herniou et al. 2013). Another nice example is the exaptation of the RNA interference and methylation-inactivation systems by eukaryotes that have now become the basis of the RNAi-based innate immunity defense mechanism of cells (Villarreal 2011).

A very interesting and well-studied innovation in mammal cells has already been introduced in chapter 4: the exaptation of the retroviral envelope protein into today's syncytins (Villarreal 2016) involved in the fusion of trophoblast cells during placenta formation could explain why embryos are protected against their mother's immune system, because the original role of these viral proteins was to interfere with host immunity. Important roles have also been suggested for a particular type of retroelement call SINE in the origin and development of the mammalian brain (Sasaki et al. 2008). One may speculate that the major differences between chimps and us derived from the differences in viral integrations that occurred upon separation of our two lineages, activating or inactivating different genes.

7.6 But ... What Is a Virus Then?

Most eukaryotic viruses share an interesting property: during infection, they build complex and specialized intracellular structures associated with membranes (sometimes in close proximity to organelles) taken from the reticulum. Inside these vesicles, viral genomes are transcribed, translated into proteins, and replicated (Romero-Brey and Bartenschlager 2014; Fernandez de Castro et al. 2016). These structures are known as "virus factories" and they provide a protected environment in which genomes can be naked while still protected from the myriad of cytoplasmic factors that cells activate as antiviral defenses to degrade them. Virus factories are comparable with obligate intracellular parasitic bacteria: they are enclosed in membranes derived from the endoplasmic reticulum, contain ribosomes and cytoskeletal elements, and recruit the mitochondria and suck ATP from it. The resemblances to cytoplasmic bacterial parasites become obvious, especially if these tiny bacteria are compared to the giant mimiviruses.

And here is the final reflection about what viruses are and whether they are alive: considering the virion particle as the true virus is equivalent to considering a grain of pollen as a redwood or an ovulum as a human. The virion would be equivalent to the germ line of the virus while the virus factory would equate to the somatic line. Following Claverie (2006), the virus factory must be considered as the real virus. With this interpretation in mind, the living nature of viruses is out of discussion. The virus factory has its own metabolism and contains ribosomes, and all information-processing processes take place isolated from the environment. Virus factories as a whole rather than as individual viral genomes will represent the living fossils of those ancestral parasites that were in the origin of the eukaryotic cell (see discussion above). So yes, we think viruses are alive and, after all, we think it is plausible that they evolved from parasitic entities that spontaneously arose coupled to pre-cellular RNA-protein replicators (hypercycles). Once membrane-bounded cells evolved, these parasites made their way into them and since then have been coevolving into what we now call viruses.

8

COMPUTER VIRUSES AND BEYOND

8.1 Viruses as Programs

In previous chapters we have considered the multiple facets of viruses, their structure and evolution, and how they explore their landscapes, strongly influencing the biosphere and our own evolution as species. But the concept of virus itself, as a parasitic entity able to propagate within and between living systems and self-perpetuating its information through a process of copying and infection, might be broader than we can imagine. With the development of a theory of cultural evolution, the possibility of defining a general form of propagating ideas, concepts, or symbols that were not supported by genetic material became a very appealing possibility. Here we want to explore the application of key concepts associated to virus dynamics to other fields and see how far the metaphors can be stretched.

As we pointed out in previous chapters, the fluid nature of genomes makes it easy for viruses to emerge. Genomes include a whole molecular toolkit that provides the raw material and rules to create a wide range of virus-like entities. This occurs under an evolutionary framework where viruses can act as replicators and spread within their worlds, pushing the boundaries of biological complexity and helping biological systems to overcome selection barriers and even acquire new properties. But there is much more than viruses. If we consider this from a more abstract perspective, we can now look at other complex systems where the nature and dynamics of viruses can provide some insights into their origins and evolution.

As philosopher Daniel Dennett points out, we can understand evolution as a process leading to a given result not very different from what computer scientists call an *algorithm*. For Dennett, Darwin's discovery is nothing but the discovery of a well-defined algorithm (Dennett 1995; Blackmore 1999). An algorithm correctly provides a formal framework to define a set of operations that, starting from a set of boundary conditions (a data set for a computer program), leads to a dynamical sequence of events (such as the operations performed by the program) that end up in some "solution" (the program output). This is one relevant piece for our discussion below, along with how successful evolutionary innovations propagate.

Three different case studies will be considered: (a) the origin and evolution of computer viruses, (b) the evolution of cancer cell populations, and (c) the emergence of human language. In each case, comparing viruses and these apparently unrelated systems will be more than an interesting set of analogies.

8.2 Emergence of Computer Viruses

In chapters 2 and 7 we have discussed the nature and inevitability of viruses within the context of digital genomes. Following von Neumann's conjecture on self-replicating machines, we have used theoretical tools to describe viruses as sort of programs that make copies of themselves using the molecular machinery of their hosts. One of the main conclusions was that parasites are the inevitable outcome of a flexible set of rules allowing the replication, cutting, and pasting (with errors) of information. A somewhat similar situation was in place a few decades after the starting of the IT revolution. Software became the driving force for IT and smaller, personal computers (figure 8.1a) became available to millions of

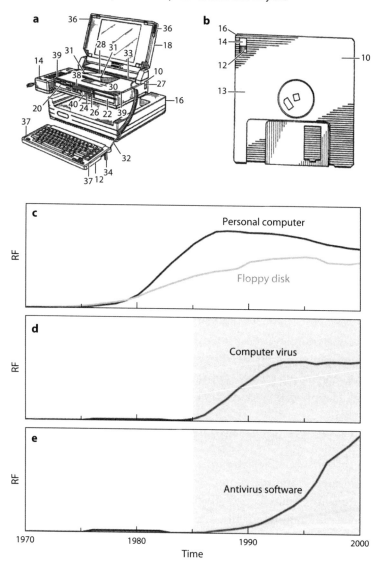

Figure 8.1. Computer viruses and the rise of information technology. Two major innovations, namely personal computers (a) and floppy disks (b) strongly contributed to the emergence and spread of computer viruses. These two innovations appear to have grown markedly (c) in the early

users, along with new magnetic storage media, particularly floppy disks (figure 8.1b). At this point, all the requirements for the propagation of computer viruses (CVs) were in place.

There are two reasons for our choice of CVs as a case study. On one hand, it gives strong support for the inevitability hypothesis: no matter whether biological or technological, the favorable preconditions lead to emergence. The second reason is that the sequence of events that took place in the evolution of CVs exhibits strong parallelisms with their biological counterparts, despite their obvious man-made (intentional) nature.[1]

Although the term *computer virus* was introduced in 1985, the beginnings of CVs can be found in 1971 (Levy 1992; Cohen 1994), when the first computer bug was created. It was limited to a simple code that copied itself on other machines using a local network of servers. The first bugs were rather harmless, and served to illustrate the principles of replication in the computer context. But virtual vandalism came shortly after. In the 1980s, a warning message like DISK ERROR on the computer screen was a sign of a coming catastrophe. And the diversity of CVs increased at a very fast pace.

The handful of CVs existing in 1990 rapidly grew, at an exponential pace. In 1996, it was estimated that more than 10,000 DOS-based CVs had been already created (Nachenberg 1997). Such growing complexity was a by-product of the

---

[1]The intentional design of CVs is a big departure from the nondesigned nature of biological evolution. However, since CVs have to coevolve with changing environments, including the design of antivirus programs, the underlying constraints on the evolution of CVs might be responsible for the universal properties shared by both.

Figure 8.1. (*Continued*) 1980s as personal computers became widely adopted by millions of users. Computer viruses (d) and antivirus software (e) followed. Data for figures c-e have been generated using the Google Ngram Viewer. Here RF = relative frequency.

response of programmers, who started to develop systems able to detect and destroy the virtual parasites. The emerging technological developments were accompanied by the use of well-known terms from epidemics. As Richard Dawkins puts it (Dawkins, 1993):

> Again predictably, the epidemic of computer viruses has triggered an arms race. Anti-viral software is doing a roaring trade. These antidote programs—"Interferon," "Vaccine," "Gatekeeper" and others—employ a diverse armory of tricks. Some are written with specific, known and named viruses in mind. Others intercept any attempt to meddle with sensitive system areas of memory and warn the user.

The list of tricks adopted by programmers in order to make their worms more effective or even invisible is too large to summarize here. At the beginning of our story, antivirus software was mainly based on detecting specific sequences of bytes that somewhat defined the signature of a given virus. Those early programs involved finding given strings by searching for them over the entire program under scrutiny. As the number of potential viruses increased, this approach became more and more involved and less efficient. New viruses became more difficult to detect and remove. Old tricks (such as appearing with a nice message such as "I love you") had to be replaced by a great invention: polymorphism (Nachenberg 1997).

Polymorphic CVs incorporated a powerful weapon mutation. As it occurs with real viruses, constantly escaping from the selection forces imposed by immune responses, detection, and removal strategies is a major driver of genomic variability. As we have already discussed, RNA viruses escape from the immune system pressure by means of an error-prone molecular machinery that is tuned to its error threshold. Polymorphic viruses are able to keep their basic functions intact and yet appear "different" to the infected machine. We can easily identify here the two types

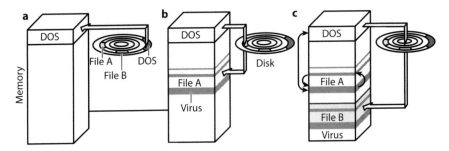

Figure 8.2 Loading the DOS operating system (a) allows a given program to be read (b) but it also provides the door for the entry of a computer virus (c). Adapted from Dewdney (1989).

of information being carried by the computer viruses: a conserved one (required for proper replication and infection) and a variable one, playing the role of the variable domains of the genome more prone to change. The new challenge was responded to by novel antivirus software, but the arms race is still unfolding.

The emergence of computer viruses constitutes a very interesting piece of cultural evolution. It illustrates quite well some key similarities between biological and artificial (designed) change, for example, the fact that mutation was the result of the conscious realization that change was a requirement. It was thus an invention, and a rather intentional one, as opposed to the intrinsic, inevitable errors that constantly take place in living systems. Moreover, we know that most mutations affecting a virus genome are harmful, impeding or threatening their replication potential. How many mutations in their man-made counterparts are lethal? None. There is a sharp separation between the part of the program associated with variation and all the rest. No interactions are allowed to occur affecting functional traits. As a consequence, the mutations are de facto simulations of mutational events, instead of a consequence of a faulty replication machinery. No true quasispecies concept can be defined here. Moreover, although the program-like nature of many viruses suggests that these entities are somehow described under the Turing machine umbrella (see chapter 2), it has been argued by various authors that Turing's formalism might fall short due to the nature of viral life cycles, which require interactions with other machines leading to spread of further programs (Thimbleby et al. 1998).

Beyond the infection and propagation dynamics of CVs there are many other aspects of their evolution that are of interest to our comparison with their biological counterparts. One in particular is the emergence of Trojans that can integrate themselves within the software of the host machines, resulting in a less harmful infection (Hruska 1990; Thimbleby et al. 1998). Being invisible to the user, not causing damage other than using some chunks of memory and perhaps processing power, they remind us of the group of lentiviruses (like HIV-1) that also integrate into the host genome and remain apparently silent over years. Trojans and other designed bugs can use computers as the sources of attack on other machines, as generators of junk information, or even as a parallel web of slave computers working together. As social networks expanded on top of computer and email webs, new threats have been emerging.

A further development in the evolutionary arms races between computer viruses and antivirus programs was the possibility of creating an artificial immune system (AIS) capable of adapting to changing infectious programs. Such a view is inspired by the natural immune responses (Farmer et al. 1986) that fight pathogens within our own bodies (Kephart et al. 1995; Forrest et al. 1997; Forrest and Beauchemin 2007). In chapter 4 we discussed some of these responses by means of simple models, particularly in the context of highly variable retroviruses. These AISs are designed in such a way that they incorporate some key aspects of immune responses, while operating as computer programs. One especially interesting component of an AIS is the potential for learning to distinguish between self and nonself (Janeway 1992; Forrest et al. 1994): as it occurs with the natural immune response, an adaptive response to CVs must identify patterns of intrusion, such as the unauthorized use of computer accounts. This requires algorithms to detect foreign activity while learning to identify stable strings of code to be protected (Forrest et al. 1994). Some mathematical results derived from the study of immune system diversity (de Boer and Perelson 1993) helped us derive theoretical bounds concerning the efficiency and cost of these AISs. The experimental design (implemented on a local computer network) confirmed the feasibility and reliable functioning of these adaptive artificial networks.

8.3 Cancer, Languages, and Minds

The surprisingly life-like features displayed by CVs are not accidental. There are universal patterns associated with viral entities, and their coevolutionary impact on their hosts (alive or man-made) gives support to this view. One especially appealing feature displayed by RNA viruses is the particular location they occupy in mutation-replication space: they live on the edge of catastrophe, right on the boundary between a phase where Darwinian evolution operates and a random phase. In the first phase, information can be preserved, while in the second one information is lost. The critical point reached by viruses illustrates a more general pattern that seems to be present in a wide variety of complex biological systems poised to criticality (Bak 1996; Solé et al. 1996; Mora and Bialek 2011; Plenz and Niebur 2014). As a consequence, some features exhibited by viral quasispecies might also emerge in other evolved contexts (Ojosnegros et al. 2011).

As we discussed in chapter 5, infectious diseases, particularly those carried by viral agents, have caused millions of deaths throughout human history. A no less lethal agent is cancer, which we also know has been around since multicellular systems emerged (Aktipis et al. 2016). Cancer is a disease that is caused by a breakdown of cell cooperation. Within our bodies, many ecological processes operate to establish a proper balance between growth and regeneration and those processes that define inhibitions and limitations to proliferation (Weinberg 2007). However, there is one kind of process that needs to be avoided: evolution. If mutated cells can escape from checkpoints, they might start growing at a faster pace and the process of Darwinian selection starts to operate. The faster the replicator, the more likely it is to win. Because of this, cancer needs to be addressed as an evolving ecosystem (Merlo et al. 2006; Spencer et al. 2006; Attolini and Michor 2009; Mas et al. 2010). With a few exceptions, cancer is not an infectious disease and thus it might appear disconnected from the world of viruses. However, there is one particular feature of most cancers that allows us to define some really interesting similarities: genetic instability.

In vivid contrast with the ordered, predictable organization of cells in healthy tissues, the observation of cancer cell populations reveals a markedly disorganized genome. It has been known since the first studies revealed that the chromosomes of these cells appear under the microscope as a gallery of horrors where all kinds of scrambled arrangements of chromosomes can be seen. How is this possible? One thing we already know from our previous discussion about how RNA viruses adapt to a changing environment is that high mutation rates provide, up to a limit, a powerful source for adaptation. Cancer cells face a somewhat similar situation. In order to avoid a plethora of control mechanisms dedicated to preventing organismal disruption, cancer cells take advantage of their evolutionary potential to escape from selection barriers (Cahill et al. 1999).

If genetic instability is used by cancer cells in ways similar to those associated to RNA virus populations, an immediate question concerns the presence of error thresholds (Solé 2003; Solé and Deisboeck 2004; Gatenby and Frieden 2007). It seems reasonable to think that small levels of instability might be an advantage: specific changes can enhance proliferation, reduce responses to inhibitions from other cells, or avoid physical constraints. But it is also reasonable to assume that too high levels of instability will necessarily trigger cell death. Is there an optimal instability rate? This idea has been suggested previously (Cahill et al. 1999) and it is worth seeing what a theoretical model can bring. In particular, we can get inspiration from the several similarities between RNA viruses and cancer cells. What type of population dynamics should be expected from a model of cancer-normal tissue competition including genetic instability? It has been suggested that unstable cancer cell populations might be also evolved close to the error catastrophe: since the failure of mechanisms that prevent instability leads to higher instability, one should expect to see tumors evolving as close to the edge as possible. If true, as discussed at the end of chapter 2, therapeutic strategies can target the Achilles heel defined by critical instability (Foz and Loeb 2010). In this context, mathematical models suggest that the nature of the instability associated to cancer cell populations (Thomas et al. 2012; Amor and Solé 2014; Solé et al. 2014) can allow the exploiting of the presence of catastrophic shifts to promote the collapse of unstable tumors.

A very different example of the relevance of quasispecies is provided by human languages. Here also we can look at their patterns of evolution and organization in terms of ecological and evolutionary dynamics (Nowak and Krakauer 1999; Christiansen and Kirby 2003; Solé et al. 2010), and multiple scales are also involved. Darwin himself considered language as some class of evolvable system and was well aware of a number of features displayed by languages that were common to species (Whitfield 2008, Solé et al. 2010). In *The Descent of Man* (Darwin 1871) Darwin explicitly says:

> The formation of different languages and of distinct species, and the proofs that both have been developed through a gradual process, are curiously parallel.

Indeed, languages exhibit properties common to species (Mufwene 2001; Pagel 2009) and exhibit an enormous diversity as illustrated by the around 6,000 languages existing today (Krauss 1992; McWhorter 2001; Nettle and Romaine 2002). Moreover, there are striking connections between genes and languages, which appear correlated at both the global (Cavalli-Sforza et al. 1988; Cavalli-Sforza 2002) and local (Lansing et al. 2007) geographic scales.

Since languages change all the time, due to a number of processes (from the generation of new words to social factors related to influence), a relevant question is, how do they preserve their internal structure and how is such large diversity maintained over long time scales? Actually, the long-term evolution of language diversity suggests that several features associated to language evolution are consistent with a quasispecies-like dynamical pattern (Nowak and Krakauer 1999; Zanette 2008; Castellano and Loreto 2009; Solé et al. 2010). The basic components of language, including both words and rules of word use, change over time. These changes can be understood from the outside (ignoring all kinds of idiosyncrasies) as some kind of "mutation." Moreover, we can conjecture that a given language is—for some reason—more easy to learn than another, thus creating the grounds for a potential definition of a fitness measure. When a model of language evolution is defined in terms of a landscape of strings, each representing a given language (each bit would be a "word"), and the flows between languages are allowed through a mutation-selection process, two main phases emerge, namely a single-language phase and a diverse language one, where multiple languages coexist. This model suggests that some dynamical properties of language evolution are not very far from those exhibited by populations of replicators.

The example of human language has been examined in many different ways. Language defines a turning point in human evolution and is one essential key for our success as a species.

But developing a theory of language capable of explaining its origins, universal properties, and evolution is still an open problem. Language has been coevolving with our brains (Deacon 1997) and is one of the essential traits that makes humans so singular (Suddendorf 2013). But language, as a nongenetic form of transmissible information, requires a process of acquisition whereby children interact with their parents. It has been suggested that, since language needs to "infect" the brains of future users, it can be seen as some sort of beneficial parasite. This picture makes sense, since language acts like a nonobligate symbiotic entity (Nerlich 1989; Christiansen 1994) that confers a clear advantage to its human hosts, without whom it cannot survive (Deacon 1997).

How good is the previous analogy? We still need to develop better models of language evolution that incorporate cognitive, cultural, and biological aspects (Christiansen and Kirby 2003). Only then will the real value of the "language as a virus" idea be properly addressed. The same must be said in relation to other uses of the term in contexts as diverse as the spread of "viral" news. As for rumors or news, it turns out that models of epidemic spreading such as those discussed in chapter 5 are very good at describing the diffusion of cultural items across social networks (Dietz 1964; Nekovee et al. 2007; Weng et al. 2012). The success of mathematical models in accounting for these processes indicates that the virus metaphor, involving some sort of agent capable of propagating through networks and infecting the underlying agents and changing their behavior, is a successful one.

REFERENCES

Abergel, C., Legendre, M., and Claverie, J. M. 2015. *The rapidly expanding universe of giant viruses: mimivirus, pandoravirus, pithovirus and mollivirus.* FEMS Microbiol. Rev. 39, 779–796.

Abrescia, N.G.A., Bamford, D. H., Grimes, J. M., and Stuart, D. I. 2012. *Structure unifies the viral universe.* Annu. Rev. Biochem. 81, 795–822.

Adami, C. 1998. *Introduction to Artificial Life.* Springer, New York.

Adami, C. 2006. *Digital genetics: unravelling the genetic basis of evolution.* Nature Reviews Genetics, 7, 109–118.

Agudelo-Romero, P., de la Iglesia, F., and Elena, S. F. 2008. *The pleiotropic cost of host-specialization in Tobacco etch potyvirus.* Infect. Genet. Evol. 8, 806–814.

Aiewsakun, P., and Katzourakis, A. 2015. *Endogenous viruses: connecting recent and ancient viral evolution.* Virology 479–480, 26–37.

Aita, T., Uchiyama, H., Inaoka, T., Nakajima, M., Kokubo, T., and Husimi, Y. 2000. *Analysis of a local fitness landscape with a model of the rough Mt. Fuji-type landscape: application to prolyl endopeptidase and thermolysis.* Biopolymers 54, 64–79.

Aktipis, C. A., Boddy, A. M., Jansen, G., et al. 2015. *Cancer across the tree of life: cooperation and cheating in multicellularity.* Phil. Trans. R. Soc. B, 370, 20140219.

Anderson, R. M., and May, R. M. 1992. *Infectious Diseases of Humans.* Oxford University Press.

Angly, F. E., Felts, B., Breitbart, M., et al. 2006. *The marine viromes of four oceanic regions.* PLoS Biol, 4, e368.

Andino, R., and Domingo, E. 2015. *Viral quasispecies.* Virology 479–480, 46–51.

Archibald, J. M. 2015. *Endosymbiosis and eukaryotic cell evolution.* Curr Biol 25, R911–921.

Arnaout, R. A., Nowak, M. A., and Wodarz, D. 2000. *HIV-1 dynamics revisited: biphasic decay by cytotoxic T lymphocyte killing?* Proc. R. Soc. B. 2267, 1347–1354.

Baltimore, D. 1971. *Expression of animal virus genomes.* Bacteriol. Rev. 35, 235–241.

Barricelli, N. A. 1962. *Numerical testing of evolution theories.* Acta Biotheoretica, 16, 69–98.

Barricelli, N. A. 1963. *Numerical testing of evolution theories.* Acta Biotheoretica, 16, 99–126.

Bedhomme, S., Lafforgue, G., and Elena, S. F. 2012. *Multihost experimental evolution of a plant RNA virus reveals local adaptation and host-specific mutations.* Mol. Biol. Evol. 29, 1481–1492.

Bell, P.J.L. 2001. *Viral eukaryogenesis: was the ancestor of the nucleus a complex DNA virus?* J. Mol. Evol. 53, 251–256.

Bell, P.J.L. 2006. *Sex and the eukaryotic cell cycle is consistent with a viral ancestry for the eukaryotic nucleus.* J. Theor. Biol. 243, 54–63.

Bennett, C. H., and Landauer, R. (1985). *The fundamental physical limits of computation.* Scientific American, 253(1), 48–56.

Bittner, B., Bonhoeffer, S., and Nowak, M. A. 1997. *Virus load and antigenic diversity.* Bull. Math. Biol. 59, 881–896.

Blackmore, S. 1999. *The meme machine.* Oxford University Press.

Bocharov, G., Ford, N. J., Edwards, J., Breinig, T., Wain-Hobson, S., and Meyerhans, A. 2005. *A genetic-algorithm approach to simulating human immunodeficiency virus evolution reveals the strong impact of multiply infected cells and recombination.* J. Gen. Virol 86, 3109–3118.

Bonhoeffer, S., Chappey, C., Parkin, N. T., Whitcomb, J. M., and Petropoulos, C. J. 2004. *Evidence for positive epistasis in HIV-1.* Science 306, 1547–1550.

Boulange, C. L., Neves, A. L., Chilloux, J., Nicholson, J. K., and Dumas, M. E. 2016. *Impact of the gut microbiota on inflammation, obesity, and metabolic disease.* Genome Med. 8, 42.

Brown, J. S., and Pavlovic, N. B. 1992. *Evolution in heterogeneous environments: effects of migration on habitat specialization.* Evol. Ecol. 6, 360–382.

Bruinsma, R. F., Gelbart, W. M., Reguera, D., Rudnick, J., and Zandi, R. 2003. *Viral self-assembly as a thermodynamic process.* Phys. Rev. Lett. 90, 248101.

Brum, J. R., and Sullivan, M. B. 2015. *Rising to the challenge: accelerated pace of discovery transforms marine virology.* Nat. Rev. Microbiol. 13, 147–159.

Bushman, F. D., Fujiwara, T., and Craigie, R. 1990. *Retroviral DNA integration directed by HIV integration protein in vitro.* Science 249, 1555–1558.

Cahill, D. P., Kinzler, K. W., Vogelstein, B., and Lengauer, C. 1999. *Genetic instability and Darwinian selection in tumours.* Trends Cell Biol. 9, M57–M60.

Cairns, J., Stent, G. S., and Watson, J. D. 2007. *Phage and the origins of molecular biology.* Cold Spring Harbor Laboratory Press. Cold Spring Harbor: USA.

Callaway, D. S., Ribeiro, R. M., and Nowak, M. A. *Virus phenotype switching and disease progression in HIV-1 infection.* Proc. R. Soc. B. 266, 2523–2530.

Case, T. J. 2000. *Illustrated Guide to Theoretical Ecology.* Oxford University Press.

Catalán, P., Arias, C. F., Cuesta, J. A., and Manrubia, S. 2017. *Adaptive multiscapes: an up-to-date metaphor to visualize molecular adaptation.* Biol. Direct 12, 7.

Cervera, H., Lalic, J., and Elena, S. F. 2016. *Effect of host species on the topography of fitness landscapes for a plant RNA virus.* J. Virol. 90, 10160–10169.

Cichutek, K., Merget, H., Norley, S., Linde, R., Kreuz, W., Gahr, M., and Kurth, R. 1992. *Development of a quasispecies of Human immunodeficiency virus type 1 in vivo.* Proc. Natl. Acad. Sci. USA 89, 7365–7369.

Clarke, D. K., Duarte, E. A., Elena, S. F., Moya, A., Domingo, E., and Holland, J. J. 1994. *The Red Queen reigns in the kingdom of RNA viruses.* Proc. Natl. Acad. Sci. USA 91, 4821–4824.

Claverie, J. M. 2006. *Viruses take center stage in cellulare evolution.* Genome Biol. 7, 110.

Clokie, M.R.J., and Mann, N. H. 2006. *Marine cyanophages and light.* Env Microbiol 8, 2074–2082.

Codoñer, F. M., Darós, J. A., Solé, R. V., and Elena, S. F. 2006. *The fittest versus the flattest: experimental confirmation of the quasispecies effect with subviral pathogens.* PLoS Pathog. 2, 1187–1193.

Coffey, L. L., Vasilakis, N., Brault, A. C., Powers, A. M., Tripet, F., and Weaver, S. C. 2008. *Arbovirus evolution in vivo is constrained by host alternation.* Proc. Natl. Acad. Sci. USA 105, 6970–6975.

Colizzi, E. S., and Hogeweg, P. 2016. *High cost enhances cooperation through the interplay between evolution and self-organisation.* BMC evolutionary biology, 16, 31.

Colson, P., De Lamballerie, X., Yutin, N., Asgari, S., et al. 2013. *"Megavirales", a proposed new order for eukaryotic nucleocytoplasmic large DNA viruses.* Arch. Virol. 158, 2517–2521.

Cooper, L. A., and Scott, T. W. 2001. *Differential evolution of eastern equine encephalitis virus populations in response to host cell type.* Genetics 157, 1403–1412.

Creager, A.N.H. *The Life of a Virus: Tobacco Mosaic Virus as an Experimental Model, 1930–1965.* University of Chicago Press.

Crick, F. 1970. *Central dogma of molecular biology.* Nature 227, 561–563.

Crick, F. H., and Watson, J. D. 1956. *Structure of small viruses.* Nature 177, 473–475.

Crick, F.H.C., and Watson, J. D. 1957. *Virus structure: general principles.* The Nature of Viruses, 5–18.

Crill, W. D., Wichman, H. A., and Bull, J. J. 2000. *Evolutionary reversals during viral adaptation to alternating hosts.* Genetics 154, 27–37.

Cuevas, J. M., Moya, A., and Elena, S. F. 2003. *Evolution of RNA virus in spatially structured heterogeneous environments.* J. Evol. Biol. 16, 456–466.

Dennett, D. C. 1995. *Darwin's dangerous idea*. The Sciences, 35, 34–40.

Desnues, C., La Scola, B., Yutin, N., Fournous, G., et al. D. 2012. *Provirophages and transpovirons as the diverse mobilome of giant viruses*. Proc. Natl. Acad. Sci. USA 109, 18078–18083.

De Visser, J.A.G.M., and Krug, J. 2014. *Empirical fitness landscapes and the predictability of evolution*. Nat. Rev. Genet. 15, 480–490.

De Visser, J.A.G.M., Hermisson, J., Wagner, G. P., Ancel Meyers, L., et al 2005. *Evolution and detection of genetic robustness*. Evolution 57, 1959–1972.

DeLong, E. F. 1997. *Marine microbial diversity: the tip of the iceberg*. Trends Biotech. 15, 203–207.

Dill, K., and Bromberg, S. 2010. *Molecular driving forces: statistical thermodynamics in biology, chemistry, physics, and nanoscience*. Garland Science.

Dolja, V. V., Kreuze, J. F., and Valkonen, J. P. 2006. *Comparative and functional genomics of Cloteroviruses*. Virus Res. 117, 38–51.

Domingo, E. 2000. *Viruses at the edge of adaptation*. Virology 270, 251–253.

Domingo, E., and Holland, J. J. 1997. *RNA virus mutations and fitness for survival*. Annu. Rev. Microbiol. 51, 151–178.

Domingo, E., Sheldon, J., and Perales, C. 2012. *Viral quasispecies evolution*. Microbiol. Mol. Biol. Rev. 76, 159–216.

Drake, J. W., Charlesworth, B., Charlesworth, D., and Crow, J. F. 1998. *Rates of spontaneous mutation*. Genetics 148, 1667–1686.

Dyson, G. 2012. *Turing's Cathedral: The Origins of the Digital Universe*. Pantheon.

Eigen, M. 1971. *Selforganization of matter and the evolution of biological macromolecules*. Naturwissenschaften 58, 465–523.

Eigen, M., McCaskill, J., and Schuster, P. 1988. *Molecular quasi-species*. J. Phys. Chem. 92, 6881–6891.

Elena, S. F., Carrasco, P., Daros, J. A., and Sanjuán, R. 2004. *Mechanisms of genetic robustness in RNA viruses*. EMBO Rep. 7, 168–173.

Elena, S. F., and Sanjuán, R. 2005. *Adaptive value of high mutation rates of RNA viruses: separating causes from consequences*. J. Virol. 79, 11555–11558.

Elena, S. F., Solé, R., and Sardanyés, J. 2010. *Simple genomes, complex interactions: epistasis in RNA virus*. Chaos 20, 026106.

Ellner, P. D. 1998. *Smallpox: gone but not forgotten*. Infection 26, 263–269.

Faria, N. R., Rambaut, A., Suchard, M. A., Baele, G., et al. 2014. *HIV epidemiology. The early spread and epidemic ignition of HIV-1 in human populations*. Science 346, 56–61.

Farnet, C. M., and Haseltine, W. A. *Integration of human immunodeficiency virus type 1 DNA in vitro*. Proc. Natl. Acad. Sci. USA 87, 4164–4168.

Fernández de Castor, I., Tenorio, R., and Risco, C. 2016. *Virus assembly factories in a lipid world*. Curr. Opin. Virol. 18, 20–26.

Flint, S. J., Enquist, L. W., Racaniello, V. R., and Skalka, A. M. 2015. *Principles of Virology*. ASM Press. Washington DC: USA.

Flores, R., Gago-Zachert, S., Serra, P., Sanjuán, R., and Elena, S.F. 2014. *Viroids: survivors from the RNA world?* Annu. Rev. Microbiol. 68, 395–414.

Forrest, S., Hofmeyr, S. A., and Somayaji, A. 1997. *Computer immunology*. Comm. ACM 40, 88–96.

Forrest, S., and Beauchemin, C. 2007. *Computer immunology*. Immunological Rev. 216, 176–197.

Forterre, P. 2002. *The origin of DNA genomes and DNA replication proteins*. Curr. Opin. Microbiol. 5, 525–532.

Forterre, P. 2006a. *The origin of viruses and their possible roles in major evolutionary transitions*. Virus Res. 117, 5–16.

Forterre, P. 2006b. *Three RNA cells for ribosomal lineages and three DNA viruses to replicate their genomes: a hypothesis from the origin of cellular domain*. Proc. Natl. Acad. Sci. USA 103, 3669–3674.

Forterre, P., and Prangishvili, D. 2009. *The origin of viruses*. Res. Microbiol. 160, 466–472.

Fraenkel-Conrat, H., and Williams, R. C. 1955. *Reconstitution of active tobacco mosaic virus from its inactive protein and nucleic acid components*. Proc. Natl. Acad. Sci USA 41, 690–698.

Freschotte, C., and Pritham, E. J. 2007. *DNA transposons and the evolution of eukaryotic genomes*. Annu. Rev. Genet. 41, 331–368.

Fry, J. D. 1996. *The evolution of host specialization: are tradeoffs overrated?* Am. Nat. 148, S84–S107.

Fuhrman, J. A. 1999. *Marine viruses and their biogeochemical and ecological effects.* Nature, 399, 541–548.

Fuhrman, J. A. 2009. *Microbial community structure and its functional implications.* Nature, 459, 193–199.

Futuyma, D. J., and Moreno, G. 1988. *The evolution of ecological specialization.* Annu. Rev. Ecol. Syst. 19, 207–233.

Murray, G. M. 1994. *The Quark and the Jaguar: Adventures in the Simple and the Complex.* Little, Brown and Co. London, UK.

Gilbert, C., Peccoud, J., Chateigner, A., Moumen, B., Cordaux, R., and Herniou, E. A. 2016. *Continuous influx of genetic material from host to virus populations.* PLoS Genet 12, e1005838.

Gilbert, W. 1986. *Origin of life: the RNA world.* Nature 319, 618.

Goldenfeld, N. 1992. *Lectures on phase transitions and the renormalization group.* Addison-Wesley.

Gotelli, N. J. 1995. *A Primer of Ecology.* Sinauer.

Greene, I. P., Wang, E., Deardorff, E. R., Milleron, R., Domingo, E., and Weaver, S. C. 2005. *Effect of alternating passage on adaptation of Sindbis virus to vertebrate and invertebrate cells.* J. Virol. 79, 14253–14260.

Hagan, M. F. 2014. *Modeling viral capsid assembly.* Adv. Chem. Phys. 155, 1–68.

Hamelaar, J. 2012. *The origin and diversity of the HIV-1 pandemic.* Trends Mol. Med. 18, 182–192.

Hannoun, C. 2013. *The evolving history of influenza viruses and influenza vaccines.* Expert. Rev. Vaccines 12, 1085–1094.

Herniou, E. A., Huguet, E., Thézé, J., Bézier, A., Periquet, G., and Drezen, J. M. 2013. *When parasitic wasps hijacked viruses: genomic and functional evolution of polydnaviruses.* Philos. Trans. R. Soc. B 368, 20130051.

Hinkley, T., Martins, J., Chappey, C., Haddad, M., et al., 2011. *A systems analysis of mutational effects in HIV-1 protease and reverse transcriptase.* Nat. Genet. 43, 487–489.

Holland, J. J., de la Torre, J. C., Clarke, D. K., and Duarte, E.A. 1991. *Quantitation of relative fitness and great adaptability of clonal populations of RNA viruses.* J. Virol. 65, 2960–2967.

Holmes, E. C. 2001. *On the origin and evolution of the human immunodeficiency virus (HIV).* Biol. Rev. Camb. Phil. Soc. 76, 239–254.

Holmes, E. C. 2010. *The RNA virus quasispecies: fact or fiction.* J. Mol. Biol. 400, 271–273.

Hopcroft, J. E. 1984. *Turing machines.* Sci. Am. 250, 86–107.

Hull, R., and Will, H. 1989. *Molecular biology of viral and nonviral retroelements.* Trends Genet. 5, 357–359.

Janeway Jr, C. A., Travers, P., Walport, M., and Shlomchik, M. J. 2004. *Immunobiology 6th edition.* Garland Science. New York.

Kaneko, K., and Ikegami, T. 1992. *Homeochaos: dynamics stability of a symbiotic network with population dynamics and evolving mutation rates.* Physica D 56, 406–429.

Kauffman, S. A. 1993. *The Origins of Order: Self-Organization and Selection in Evolution.* Oxford University Press.

Kauffman, S., and Levin, S. 1987. *Towards a general theory of adaptive walks on rugged landscapes.* J. Theor. Biol. 128, 11–45.

Kawecki, T. J. 1994. *Accumulation of deleterious mutations and the evolutionary cost of being a generalist.* Am. Nat. 144, 833–838.

Kawecki, T. J. 2000. *Adaptation to marginal habitats: contrasting influence of the dispersal rate on the fate of alleles with small and large effects.* Proc. R. Soc. B 267, 1315–1320.

Keeling, M. J., and Eames, K. T. 2005. *Networks and epidemic models.* J. R. Soc. Interface 2, 295–307.

Keeling, M. J., and Rohani, P. 2007. *Modeling Infectious Disease in Humans and Animals.* Princeton University Press.

Kelly, S. G., and Taiwo, B. O. 2015. *HIV reservoirs in lymph nodes and spleen.* Encyclopedia of AIDS. DOI 10.1007/978.

Kerr, P. J., Liu, J., Cattadori, I., Ghedin, E., Read, A. F., and Holmes, E. C. 2015. *Myxoma virus and the Leporiposviruses: an evolutionary paradigm.* Viruses 7, 1020–1061.

Kingman, J. 1987. *A simple model for the balance between selection and mutation.* J. Appl. Probab. 15, 1–12.

Kloverpris, H. N., Leslie, A., and Goulder, P. 2016. *Role of HLA adaptation in HIV evolution.* Front. Immunol. 6, 665.

Klug, A. 1999. *The tobacco mosaic virus particle: structure and assembly.* Phil. Trans. R. Soc. B 354, 531–535.

Koelle, K., Cobey, S., Grenfell, B. T., and Pascual, M. 2006. *Epocal evolution shapes the phylodynamics of inerpandemic influenza A (H3N2) in humans.* Science 314, 1898–1903.

Koonin, E. V., and Dolja, V. V. 2013. *A virocentric perspective on the evolution of life.* Curr. Opin. Virol. 3, 546–557.

Koonin, E. V., Senkevich, T. G., and Dolja, V. V. 2006. *The ancient virus world and evolution of cells.* Biol. Direct 1, 29.

Koonin, E. V., and Starokadomskyy, P. 2016. *Are viruses alive? The replicator paradigm sheds decisive light on an old but misguided question.* Studies in History and Philosophy of Biological and Biomedical Sciences, 59, 125–134.

Kouyos, R. D., Leventhal, G. E., Hinkley, T., Haddad, M., et al. 2012. *Exploring the complexity of the HVI-1 fitness landscape.* PLoS Genet. 8, e1002551.

Kristensen, D. M., Mushegian, A. R., Dolja, V. V., and Koonin, E. V. 2010. *New dimensions of the virus world discovered through metagenomics.* Trends Microbiol. 18, 11–19.

Krupovic, M., and Bamford, D. H. 2010. *Order to the viral universe.* J. Virol. 84, 12476–12479.

Krupovic, M., and Koonin, E.V. 2016. *Self-synthesizing transposons: unexpected key players in the evolution of viruses and defense systems.* Curr. Opin. Microbiol. 30, 25–33.

Kushner, D. J. 1969. *Self-assembly of biological structures.* Bacteriological reviews, 33, 302–345.

Lalic, J., and Elena, S. F. 2012. *Magnitude and sign epistasis among deleterious mutations in a positive-sense plant RNA virus.* Heredity 109, 71–77.

Lalic, J., and Elena, S. F. 2015. *The impact of higher-order epistasis in the within-host fitness of a positive-sense plant RNA virus.* J. Evol. Biol. 28, 2236–2247.

Lane, J. M. 2006. *Mass vaccination and surveillance/containment in the eradication of smallpox.* Curr. Top. Microbiol. Immunol. 304, 17–29.

La Scola, B., Audic, S., Robert, C., et al. 2003. *A giant virus in amoebae.* Science, 299, 2033–2033.

La Scola, B., Desnues, C., Pagnier, I., Robert, C., et al. 2008. *The virophage as a unique parasite of the giant mimivirus.* Nature 455, 100–104.

Lauring, A. S., and Andino, R. 2010. *Quasispecies theory and the behavior of RNA viruses.* PLoS Pathog. 6, e1001005.

Leuthäusser, I. 1986. *An exact correspondence between Eigen's evolution model and a two-dimensional Ising system.* J. Chem. Phys. 84, 1884–1885.

Leuthäusser, I. 1987. *Statistical mechanics of Eigen's evolution model.* J. Stat. Phys. 48, 343–360.

Levin, S. A. 1998. *Ecosystems and the biosphere as complex adaptive systems.* Ecosystems 1, 431–436.

Li, C. X., Shi, M., Tian, J. H., Lin, X. D., et al. 2015. *Unprecedent genomic diversity of RNA viruses in arthropods reveals the ancestry of negative-sense RNA viruses.* eLIFE 4, e05378.

Littlejohn, M., Locarnini, S., and Yuen, L. 2016. *Origins and evolution of Hepatitis B virus and hepatitis delta virus.* Cold Spring Harb. Perspect. Med. 6, a021360.

Lively, C. M. 2010. *A review of Red Queen models for the persistence of obligate sexual reproduction.* J. Hered. 101, S13–S20.

Loeb, L. A., and Mullins, J. I. 2000. *Lethal Mutagenesis of HIV by Mutagenic Ribonucleoside Analogs.* AIDS research and human retroviruses, 16(1), 1–3.

López-Garcia, P., and Moreira, D. 2009. *Yet viruses cannot be included in the three of life.* Nat. Rev. Microbiol. 7, 615.

Lorenzo-Redondo, R., Fryer, H. R., Bedford, T., et al. 2016. *Persistent HIV-1 replication maintains the tissue reservoir during therapy.* Nature 530, 51–56.

Lynch, M., and Conery, J. S. 2003. *The origins of genome complexity.* Science 302, 1401–1404.

May, R. M. 2001. *Stability and Complexity in Model Ecosystems.* Princeton University Press.

May, R. M. 2004. *Uses and abuses of mathematics in biology.* Science 303, 790–793.

Maynard Smith, J. 2000. *The concept of information in biology.* Philosophy of science, 67, 177–194.

Mills, D. R., Peterson, R. L., and Spiegelman, S. 1967. *An extracellular Darwinian experiment with a self-duplicating nucleic acid molecule.* Proc. Natl. Acad. Sci. USA 58, 217–224.

Minot, S., Bryson, A., Chehoud, C., Wu, G. D., Lewis, J. D., and Bushman, F. D. (2013). *Rapid evolution of the human gut virome.* Proc. Natl. Acad. Sci. USA 110, 12450–12455.

Mitchell, M. 2009. *Complexity: A Guided Tour*. Oxford University Press.

Molina-París, C., and Lythe, G. (Eds.). 2011. *Mathematical models and immune cell biology*. Springer.

Morange, M. 2000. *A History of Molecular Biology*. Harvard University Press.

Motulsky, A. G. 1964. *Hereditary red cell traits and malaria*. Am. J. Trop. Med. Hyg. 13, S147–S158.

Mouritsen, O. G. 2005. *Life—As a Matter of Fat*. Springer-Verlag.

Mushegian, A. R., and Elena, S. F. 2015. *Evolution of plant virus movement proteins from the 30K superfamily and of their homologs integrated in plant genomes*. Virology 476, 304–315.

Myllykallio, H., Lipowski, G., Leduc, D., Fillon, J., Forterre, P., and Liebl, U. 2002. *An alternative flavin-dependent mechanism for thymidylate synthesis*. Science 297, 105–107.

Nachenberg, C. 1997. *Computer virus-coevolution*. Comm. ACM 50, 46–51.

Novella, I. S., Duarte, E. A., Elena, S. F., Moya, A., Domingo, E., and Holland, J. J. 1995. *Exponential fitness increases of RNA virus fitness during large population transmissions*. Proc. Natl. Acad. Sci. USA 92, 5841–5844.

Novella, I. S., Hershey, C. L., Escarmis, C., Domingo, E., and Holland, J. J. 1999. *Lack of evolutionary stasis during alternating replication of an arbovirus in insect and mammalian cells*. J. Mol. Biol. 287, 459–465.

Nowak, M. A., Anderson, R. M., Boerlijst, M. C., Bonhoeffer, S., May, R. M., and McMichael, A. J. 1996. *HIV-1 evolution and disease progression*. Science 274, 1008–1011.

Nowak, M. A., Anderson, R. M., McLean, A. R., Wolfs, T. F., Goudsmit, J., and May, R. M. 1991. *Antigenic diversity thresholds and the development of AIDS*. Science 254, 963–969.

Nowak, M. A., and Bangham, C. R. *Population dynamics of immune responses to persistent viruses*. Science 272, 74–79.

Nowak, M., and May, R. M. 2000. *Virus Dynamics: Mathematical Principles of Immunology and Virology*. Oxford University Press, UK.

Palukaitis, P. 2016. *Satellite RNAs and satellite viruses*. Mol. Plant Microb. Interact. 29, 181–186.

Pantaleo, G., Graziosi, C., and Fauci, A. S. 1993. *The immunopathogenesis of human immunodeficiency virus infection.* New England Journal of Medicine 328, 327–335.

Perales, C., Iranzo, J., Manrubia, S. C., and Domingo, E. 2012. *The impact of quasispecies dynamics on the use of therapeutics.* Trends Microbiol. 20, 595–603.

Perelson, A. S., and Kauffman, S. A. (eds.) 1991. *Molecular Evolution on Rugged Landscapes.* Addison-Wesley.

Perelson, A. S., and Weisbuch, G. 1997. *Immunology for physicists.* Reviews of modern physics 69, 1219.

Perelson, A. S., and Nelson, P. W. 1999. *Mathematical analysis of HIV-1 dynamics in vivo.* SIAM review, 41, 3–44.

Perelson, A. S. 2002. *Modelling viral and immune system dynamics.* Nature Rev. Immunol. 2, 28–36.

Perrin, P. 2015. *Human and tuberculosis coevolution: an integrative view.* Tuberculosis 95, S112–S116.

Philippe, N., Legendre, M., Doutre, G., et al. 2013. *Pandoraviruses: amoeba viruses with genomes up to 2.5 Mb reaching that of parasitic eukaryotes.* Science 341, 281–286.

Presloid, J. B., Ebendick-Corpus, B. E., Zárate, S., and Novella, I. S. 2008. *Antagonistic pleiotropy involving promoter sequences in a virus.* J. Mol. Biol. 382, 342–352.

Quer, J., Huerta, R., Novella, I. S., Tsimring, L. S., Domingo, E., and Holland, J. J. 1996. *Reproducible nonlinear population dynamics and critical points during replicate competitions of RNA virus quasispecies.* J. Mol. Biol. 264, 465–471.

Rambaut, A., Roberston, D. L., Pybus, O. G., Peeters, M., and Holmes, E. C. 2001. *Human immunodeficiency virus. Phylogeny and the origin of HIV-1.* Nature 410, 1047–1048.

Raoult, D., Audic, S., Robert, C., Abergel, C., et al. 2004. *The 1.2-megabase genome sequence of Mimivirus.* Science 306, 1344–1350.

Ray T. S. 1991. An approach to the synthesis of life. 371–408. In: *Artificial life II: Santa Fe Institute studies in the sciences of complexity.* Langton, C., Taylor, C. and Farmer, D. (eds). Addison-Wesley.

Ray, T. S. 1994. *Evolution, complexity, entropy and artificial reality.* Physica D 75, 239–263.

Reidys, C. M., and Stadler, P. F. 2002. *Combinatorial landscapes.* SIAM review, 44, 3–54.

Remold, S. K., Rambaut, A., and Turner, P. E. 2008. *Evolutionary genomics of host adaptation in vesicular stomatitis virus.* Mol. Biol. Evol. 25, 1138–1147.

Rico, P., Ivars, P., Elena, S. F., and Hernandez, C. 2006. *Insights into the selective pressures restricting Pelargonium flower break virus genome variability: evidence for host adaptation.* J. Virol. 80, 8124–8132.

Romero-Brey, I., and Bartenschlager, R. 2014. *Membraneous replication factories induced by plus-strand RNA viruses.* Viruses 6, 2826–2857.

Roossinck, M. J., Martin, D. P., and Roumagnac, P. 2015. *Plant virus metagenomics: advances in virus discovery.* Phytopathology 105, 716–727.

Rossmann, M. G., and Rao, V. B. 2012. *Viruses: sophisticated biological machines.* In: *Viral Molecular Machines* (pp. 1-3). Springer.

Ryan, F. P. 2016. *Viral symbiosis and the holobiontic nature of the human genome.* APMIS 124, 11–19.

Sanger, F., Air, G. M., Barrell, B. G., et al. 1977. *Nucleotide sequence of bacteriophage φX174 DNA.* Nature, 265, 687–695.

Sanjuán, R., Codoñer, F. M., Moya, A., and Elena, S. F. 2004. *Natural selection and the organ-specific differentiation of HIV-1 V3 hypervariable region.* Evolution 58, 1185–1194.

Sanjuán, R., Cuevas, J. M., et al. 2009. *Selection for robustness in mutagenized RNA viruses.* PLoS Genet. 3, 939–946.

Sanjuán, R., Moya, A., and Elena, S. F. 2004. *The contribution of epistasis to the architecture of fitness in an RNA virus.* Proc. Natl. Acad. Sci. USA 101, 15376–15379.

Sanjuán, R., Nebot, M. R., Chirico, N., Mansky, L. M., and Belshaw, R. 2010. *Viral mutation rates.* J. Virol. 84, 9733–9748.

Sardanyés, J., Elena, S. F., and Solé, R. V. 2008. *Simple quasispecies models for the survival-of-the-flattest effect: the role of space.* J. Theor. Biol. 250, 560–568.

Sasaki, T., Nishihara, H., Hirakawa, M., Fujimura, K., et al. 2008. *Possible involvement of SINEs in mammalian-specific brain formation.* Proc. Natl. Acad. Sci. USA 105, 4220–4225.

Schuster, P., and Swetina, J. 1988. *Stationary mutant distributions and evolutionary optimization*. Bull. Math. Biol. 50, 635–660.

Schuster, P. 2006. *Prediction of RNA secondary structures: from theory to models and real molecules*. Reports on Progress in Physics, 69, 1419.

Schuster, P. 2009. *Genotypes and phenotypes in the evolution of molecules*. European Review, 17, 281–319.

Serrao, E., and Engelman, A. N. 2016. *Sites of retroviral DNA integration: from basic research to clinical applications*. Crit. Rev. Biochem. Mol. Biol. 51, 26–42.

Simmons, P. 2015. *Methods for virus classification nd the challenge of incorporating metagenomic sequence data*. J. Gen. Virol. 96, 1193–1206.

Smyth, R. P., Davenport, M. P., and Mak, J. 2012. The origin of genetic diversity in HIV-1. Virus research 169 415–429.

Solé, R. V., Ferrer, R., Gonzalez-Garcia, I., Quer, J., and Domingo, E. 1999. *Red Queen dynamics, competition and critical points in a model of RNA virus quasispecies*. J. Theor. Biol. 198, 47–59.

Solé, R. V., and Goodwin, B. 2002. *Signs of Life: How Complexity Pervades Biology*. Basic Books.

Solé, R. 2011. *Phase Transitions*. Princeton University Press.

Spiegelman, S. 1971. *An aproach to the experimental analysis of precellular evolution 1*. Quarterly reviews of biophysics, 4, 213–253.

Stadler, P. F. 1999. *Fitness landscapes arising from the sequence-structure maps of biopolymers*. Journal of Molecular Structure 463, 7–19.

Stadler, B. M., Stadler, P. F., Wagner, G. P., and Fontana, W. 2001. *The topology of the possible: Formal spaces underlying patterns of evolutionary change*. J. Theor. Biol. 213, 241–274.

Stafford, M. A., Corey, L., Cao, Y., et al. 2000. *Modeling plasma virus concentration during primary HIV infection*. J. Theor Biol. 203, 285–301.

Stewart, I. 1998. *Life's Other Secret. The New Mathematics of the Living World*. John Wiley.

Stockley, P. G., Twarock, R., Bakker, S. E., Baker, A. M., et al. 2013. *Packaging signals in single-stranded RNA viruses: nature's alternative to a purely electrostatic assembly mechanism*. J. Biol. Phys. 39, 277–287.

Suttle, C. A. 2005. *Viruses in the sea*. Nature, 437, 356–361.

Szathmáry, E. 2006. *The origin of replicators and reproducers.* Phil. Trans. Roy. Soc. B 361, 1761–1776.

Thomas, C. M., and Summers, D. 2008. *Bacterial plasmids.* Encyclopedia of Life Sciences, DOI: 10.1002/9780470015902.a0000468.pub2.

Taur, Y., and Pamer, E. G. 2016. *Microbiome mediation of infections in the cancer setting.* Genome Med. 8, 40.

Tripathi, K., Balagam, R., Vishnoi, N. K., and Dixit, N. M. 2012. *Stochastic simulations suggest that HIV-1 survives close to its error threshold.* PLoS Comput Biol, 8, e1002684.

Turner, P. E., and Elena, S. F. 2000. *Cost of host radiation in an RNA virus.* Genetics 156, 1465–1470.

Van Nimwegen, E. 2006. *Influenza escapes immunity along neutral networks.* Science 314, 1884–1886.

Van Nimwegen, E., Crutchfield, J. P., and Huynen, M. (1999). *Neutral evolution of mutational robustness.* Proc. Natl. Acad. Sci. USA 96, 9716–9720.

Van Tienderen, P. H. 1991. *The evolution of generalists and specialists in spatially structured heterogeneous environments.* Evolution 45, 1317–1331.

Villarreal, L. P. 2004. *Are viruses alive?* Scientific American 97, 102.

Villarreal, L. P. 2011. *Viral ancestors of antiviral systems.* Viruses 3, 1933–1958.

Villarreal, L. P. 2016. *Viruses and the placenta: the essential virus first view.* APMIS 124, 20–30.

De Visser, J.A.G., and Krug, J. 2014. *Empirical fitness landscapes and the predictability of evolution.* Nature Reviews Genetics 15, 480–490.

Von Neumann, J., and Burks, A. W. 1966. *Theory of self-reproducing automata.* IEEE Transactions on Neural Networks, 5, 3–14.

Wagner, P. L., and Waldor, M. K. 2002. *Bacteriophage control of bacterial virulence.* Infect. Immun. 70, 3985–3993.

Weaver, S. C., Brault, A. C., Kang, W., and Holland, J. J. 1999. *Genetic and fitness changes accompanying adaptation of an arbovirus to vertebrate and invertebrate cells.* J. Virol. 73, 4316–4326.

Weinberg, R. 2013. *The Biology of Cancer.* Garland Science.

Wesemann, D. R., and Nagler, C. R. 2016. *The microbiome, timing, and barrier function in the context of allergic disease.* Immunity 44, 728–738.

Whitlock, M. C. 1996. *The Red Queen beats the jack-of-all-trades: the limitations on the evolution of phenotypic plasticity and niche breadth.* Am. Nat. 148, S65–S77.

Wilke, C. O. 2001. *Selection for fitness versus selection for robustness in RNA secondary structure folding.* Evolution 55, 2412–2420.

Wilke, C. O., Forster, R., and Novella, I. S. 2006. *Quasispecies in time-dependent environments.* Curr. Top. Microbiol. Immunol. 299, 33–50.

Willard-Mack, C. L. 2006. *Normal structure, function, and histology of lymph nodes.* Toxicol. Pathol. 34, 409–424.

Wodarz, D., and Nowak, M. A. 1999. *Evolutionary dynamics of HIV-induced subversion of the immune response.* Immunol. Rev. 168, 75–89.

Wodarz, D., and Nowak, M.A. 2002. *Mathematical models of HIV pathogenesis and treatment.* Bioessays 24, 1178–1187.

Woolhouse, M.E.J., and Gowtage-Sequeria, S. 2005. *Host range and emerging and reemerging pathogens.* Emerging Infect. Dis. 11, 1842–1847.

Woolhouse, M.E.J., Taylor, L. H., and Haydon, D. T. 2001. *Population biology of multihost pathogens.* Science 292, 1109–1112.

Yoon, S. W., Webby, R. J., and Webster, R. G. 2014. *Evolution and ecology of influenza A viruses.* Curr. Top Microbiol. Immunol. 385, 359–375.

Zandi, R., Reguera, D., Bruinsma, R. F., Gelbart, W. M., and Rudnick, J. 2004. *Origin of icosahedral symmetry in viruses.* Proc. Natl. Acad. Sci. USA 101, 15556–15560.

Zhou, J., Zhang, W., Yan, S., Xiao, J., Zhang, Y., Li, B., Pan, Y., and Wang, Y. 2013. *Diversity of virophages in metagenomic data sets.* J. Virol. 87, 4225–4236.

INDEX